Laurent Balet

Coupling of quantum dots to photonic crystal nanocavities

Laurent Balet

Coupling of quantum dots to photonic crystal nanocavities

Investigation into the coupling of quantum dots to photonic crystal nanocavities at telecommunication wavelengths

Südwestdeutscher Verlag für Hochschulschriften

Impressum/Imprint (nur für Deutschland/ only for Germany)
Bibliografische Information der Deutschen Nationalbibliothek: Die Deutsche Nationalbibliothek verzeichnet diese Publikation in der Deutschen Nationalbibliografie; detaillierte bibliografische Daten sind im Internet über http://dnb.d-nb.de abrufbar.
Alle in diesem Buch genannten Marken und Produktnamen unterliegen warenzeichen-, marken- oder patentrechtlichem Schutz bzw. sind Warenzeichen oder eingetragene Warenzeichen der jeweiligen Inhaber. Die Wiedergabe von Marken, Produktnamen, Gebrauchsnamen, Handelsnamen, Warenbezeichnungen u.s.w. in diesem Werk berechtigt auch ohne besondere Kennzeichnung nicht zu der Annahme, dass solche Namen im Sinne der Warenzeichen- und Markenschutzgesetzgebung als frei zu betrachten wären und daher von jedermann benutzt werden dürften.

Verlag: Südwestdeutscher Verlag für Hochschulschriften Aktiengesellschaft & Co. KG
Dudweiler Landstr. 99, 66123 Saarbrücken, Deutschland
Telefon +49 681 37 20 271-1, Telefax +49 681 37 20 271-0, Email: info@svh-verlag.de
Zugl.: Lausanne, EPFL, Diss., 2009

Herstellung in Deutschland:
Schaltungsdienst Lange o.H.G., Berlin
Books on Demand GmbH, Norderstedt
Reha GmbH, Saarbrücken
Amazon Distribution GmbH, Leipzig
ISBN: 978-3-8381-0960-2

Imprint (only for USA, GB)
Bibliographic information published by the Deutsche Nationalbibliothek: The Deutsche Nationalbibliothek lists this publication in the Deutsche Nationalbibliografie; detailed bibliographic data are available in the Internet at http://dnb.d-nb.de.
Any brand names and product names mentioned in this book are subject to trademark, brand or patent protection and are trademarks or registered trademarks of their respective holders. The use of brand names, product names, common names, trade names, product descriptions etc. even without a particular marking in this works is in no way to be construed to mean that such names may be regarded as unrestricted in respect of trademark and brand protection legislation and could thus be used by anyone.

Publisher:
Südwestdeutscher Verlag für Hochschulschriften Aktiengesellschaft & Co. KG
Dudweiler Landstr. 99, 66123 Saarbrücken, Germany
Phone +49 681 37 20 271-1, Fax +49 681 37 20 271-0, Email: info@svh-verlag.de

Copyright © 2009 by the author and Südwestdeutscher Verlag für Hochschulschriften Aktiengesellschaft & Co. KG and licensors
All rights reserved. Saarbrücken 2009

Printed in the U.S.A.
Printed in the U.K. by (see last page)
ISBN: 978-3-8381-0960-2

Contents

Abstract   5

Kurzfassung   7

Résumé   9

1 Introduction   11
- 1.1 Single-photon sources 12
  - 1.1.1 Motivation for single-photon sources 12
  - 1.1.2 Various single-photon emitters 14
  - 1.1.3 Self-assembled QDs with emission around $1.3\,\mu m$ 16
- 1.2 Purcell effect 20
  - 1.2.1 The Purcell factor 21
  - 1.2.2 Corrections for Quantum Dots 23
  - 1.2.3 Coupling efficiency 25
- 1.3 Photonic Crystals 25
  - 1.3.1 The master equation 25
  - 1.3.2 Two-dimensional photonic crystals 27
  - 1.3.3 Two-dimensional photonic crystals in membrane 28
  - 1.3.4 Photonic crystal cavities 29
- 1.4 Outline of the manuscript 31

2 Experimental techniques   35
- 2.1 Photonic crystal fabrication 35
- 2.2 Measurement setups 42
  - 2.2.1 Micro-photoluminescence setup 42
  - 2.2.2 Superconducting Single-Photon Detector 43
  - 2.2.3 Cryogenic probestation 44
  - 2.2.4 Tri-axial micro-photoluminescence setup 45

3	**Tuning**	**47**
	3.1 Lithographic tuning	49
	3.2 Temperature tuning	56
	3.3 Gas deposition at cryogenic temperatures	59
	3.4 Laser heating and thermal annealing	61
	3.5 Global and local infiltration with polymers	64
	3.6 SNOM spectral tuning	68
	3.7 SNOM local heating	71
	3.8 Double membrane tuning	75
	3.9 Electric field tuning	94
4	**Harvesting of light**	**97**
	4.1 Vertical extraction	97
	4.1.1 Design	98
	4.1.2 Experimental realization	100
	4.2 Cavity in waveguides	104
	4.2.1 Design	104
	4.2.2 Experiment	105
5	**Control of the spontaneous emission rate**	**109**
	5.1 Single Quantum Dots in a Photonic Crystal nanocavity	109
	5.1.1 Experimental method	109
	5.1.2 Detuning	111
	5.1.3 Time-resolved measurements	112
	5.1.4 Detuning and effective Purcell factor	114
	5.1.5 Conclusion	115
	5.2 Photonic crystal LED	115
	5.2.1 Device fabrication	116
	5.2.2 Electroluminescence under continuous bias operation	118
	5.2.3 Time-resolved electroluminescence	120
	5.2.4 Conclusion	121
6	**Conclusion**	**123**

A	Material dispersion in the master equation	125
B	Modes of selected photonic crystal cavities	127
	B.1 H1 cavity	128
	B.2 L3 cavity	129
	B.3 Modified L3 cavity	131

Bibliography   131

Acknowledgements   151

Abstract

Recently, the emission of single photons with emission wavelength in the 1.3 μm telecommunication window was demonstrated for InAs quantum dots. This makes them strong candidates for applications such as quantum cryptography, and in a longer term, quantum computing. However, efficient extraction of the spontaneous emission from semiconductors still represents a major challenge due to total internal reflection at the semiconductor/air interface. In particular, single photon sources based on quantum dots are plagued by low extraction efficiency and poor coupling to single-mode fibers, typically on the order of $10^{-3} \sim 10^{-4}$, which prevents their application to quantum communication.

To seek a solution to this problem, this thesis work explores the integration of quantum dots, with emission at 1.3 μm, in photonic crystal microcavities. Photons emitted in a mode of the cavity are funneled out of the semiconductor, and thus bypass the total internal reflection. In addition, the modified density of electromagnetic states in the cavity affects the emission lifetime of a weakly coupled emitter: in resonance, we assist to an increase of the emission rate, known as the Purcell effect, that would allow faster data transmission. Photonic crystal microcavities conveniently address this objective as they provide modes with the required small volumes and high quality factors. They also allow the engineering of the farfield pattern of the cavity modes, and thus of the collection efficiency.

In the following pages, after briefly reviewing single photon emitters, the Purcell effect, and photonic crystal cavities, we present our results on the coupling of quantum dots to photonic crystal cavities. We report on the different strategies we used to control the tuning between the cavity mode and the quantum dot emission frequency. We also show our efforts in improving the collection of coupled photons by engineering the shape of the microcavity. Finally, we present our time-resolved measurements demonstrating the Purcell effect under optical and electrical operation.

Keywords: semiconductor, quantum dot, photonic crystal, microcavity, Purcell effect, light emitting diode (LED), micro-photoluminescence, time-resolved spectroscopy

Kurzfassung

InAS-Quantenpunkte können Einzel-Photonen mit Wellenlängen im Bereich von 1.3 µm emittieren. Da diese Wellenlänge in der Telekommunikation verwendet wird, bieten sich Anwendungsmöglichkeiten in der Quantenkryptographie und in weiter Zukunft in Quanten-Rechnern. Jedoch stellt die effiziente Extraktion der spontanen Lichtemission von Halbleitern eine Herausforderung dar, da es an der Grenzschicht Halbleiter - Luft zu Reflexionen kommt. Insbesondere Ein-Photonen-Quellen leiden unter geringen Extraktionskoeffizienten ($10^{-3} \sim 10^{-4}$) und schwacher Kopplung mit Monomodfasern. Dies unterbindet ihre Anwendung in der Telekommunikation bis jetzt.

Mit dem Ziel die Extraktionskoeffizienten und die Kopplung mit Monomodfasern zu verbessern, werden in dieser Arbeit Quantenpunkte mit Emissionswellenlängen von 1.3 µm in photonische Kristalle mit Punktdefekten integriert. Photonen, die in der Mode des photonischen Kristalls emittiert werden, überwinden die interne Reflexion und werden gerichtet aus dem Halbleiter emittiert. Zusätzlich beeinflusst der elektromagnetische Zustand des photonischen Kristalls die Emissionsdauer von schwach gekoppelten Strahlern. Im resonanten Mode wird eine erhöhte Emissionsrate, der so genannte Purcell-Effekt, beobachtet. Dies ermöglicht höhere Übertragungsraten. Photonische Kristalle 1-Moden Resonatoren können Licht in kleinsten Volumina einschliessen und erzielen gute Qualitätsfaktoren. Zudem lässt sich das Fernfeld der Ausgangsmode (Far-Field Pattern) modulieren, was zu einem effizienteren Einsammeln der Photonen führt.

Am Anfang der Arbeit wird ein kurzer Überblick über Ein-Photonen-Quellen, den Purcell-Effekt und photonische Kristall Resonatoren gegeben. Anschliessend werden die durch Integration von Quantenpunkten in photonische Kristalle erzielten Ergebnisse präsentiert. Gezeigt werden angewendete Strategien um den Mode des photonischen Kristalls gezielt mit der Emissionsfrequenz des Quantenpunktes einzustimmen. Ein verbessertes Einsammelm von gekoppelten Photonen wurde durch strukturelle Veränderungen des photonischen Kristalls erreicht. Zeitaufgelöste Messungen unter optischen und elektrischen Anregung haben es erlaubt den Purcell-Effekt nachzuweisen.

Stichwörter: Halbleiter, Quantenpunkt, photonische Kristall Resonatoren, Purcell-

Effekt, lichtemittierende Diode (LED), Mikrophotolumineszenz

Résumé

Depuis peu, le fonctionnement en régime de photon unique a été démontré pour des boîtes quantiques InAs avec une longueur d'onde d'émission à 1.3 μm. Ce sont donc des candidats particulièrement intéressants pour des applications telles que la cryptographie par chiffrement quantique ou, à plus long terme, pour réaliser un calculateur quantique. Cependant, l'extraction de la lumière produite au cœur d'un semi-conducteur est entravée par la réflexion totale interne à l'interface avec l'air. En particulier, pour les boîtes quantiques mentionnée ci-dessus, l'efficacité d'extraction et de couplage en fibre optique des photons est de l'ordre de 1‰ à 0.1‰. Ainsi, même si leur longueur d'onde les rend attractifs pour les télécommunications par fibre optique, le faible rendement et sa nature aléatoire limite leur efficacité pour les applications quantiques.

L'objet de ce travail de thèse est d'étudier une solution à ce problème en couplant les boîtes quantiques avec des micro-cavités à cristal photonique. En effet, le mode de cavité agit comme un canal qui permet de guider les photons hors du semi-conducteur, réduisant ainsi les problèmes de réflexion à l'interface. En outre, la densité d'états électromagnétiques modifiée par l'effet de cavité va agir sur le temps de vie de l'émetteur faiblement couplé : en résonance, le taux d'émission se voit augmenter. Cet effet Purcell, d'après le nom de son découvreur, est bénéfique pour les communications, puisqu'il permet d'accroître la vitesse du transfert de données. Les cavités à cristaux photoniques sont particulièrement intéressantes car elles possèdent le faible volume et le grand facteur de qualité requis pour l'observation de l'effet Purcell. De plus, en ajustant judicieusement leur forme, elles permettent de modeler le champ lointain d'émission de la cavité, et donc d'optimiser la collection des photons par un système optique.

Pour commencer, nous passons brièvement en revue les sources à photon uniques, l'effet Purcell et les cavités à cristaux photoniques. Ensuite, nous présentons les différentes stratégies que nous avons utilisées pour amener le mode de cavité en résonance avec l'émission des boîtes. Nous montrerons aussi notre approche d'optimisation de l'efficacité de collection en retouchant légèrement la forme de la cavité. Enfin, nos mesures, sous pompage optique et électrique, de la dynamique d'émission des boîtes quantiques couplées à une micro-cavité, nous permettent de mettre en évidence l'effet Purcell.

Mots-clés : semiconducteur, boîte quantique, cristal photonique, microcavité, effet Purcell, diode électroluminescente, LED, micro-photoluminescence, spectroscopie résolue en temps

1

Introduction

Nowadays, the dimension of single memory cells in a computer microprocessors is on the order of a few tens of nanometers[a]. Further scaling down will eventually bring the individual constituents of such processors to the atomic length scale. On the other hand, some computing tasks, like the factorization of large numbers, could be solved more efficiently with algorithms based on the laws of quantum mechanics. One possible pathway to the realization of quantum computers is through the development of optical quantum information processing. Photons will then be used not only to transmit information, but to perform logical operations, in a similar way as the electrons in an electrical circuit.

To achieve this, we need light sources able to produce a well defined number of photons on demand. We also need a way to manipulate and control the propagation of those photons. This can be done with photonic crystals, which can be seen as a "semiconductor for light". They can be used to slow down the light, to make sharp bends with low losses, thus reducing the dimensions of the device, and to build cavities with volume as small as a cubic wavelength of light.

The road to create such quantum computers is still long. However, the research in this area already led to exciting realizations, the most famous being quantum cryptography: an intrinsically secure communication channel by mean of quantum key distribution.

In this thesis, I investigate the integration of quantum dots, a particular type of single photon emitter, into a photonic crystal microcavity. The cavity enhances the properties

---

[a] Commercially available Intel Core i7: 45 nm (2008)

of the light source by increasing the emission rate, thus allowing for faster data transfer. I operate in a wavelengths range near 1.3 µm in the near-infrared, as this corresponds to a telecommunication window with low absorption in optical fibers. Working with single photons in this wavelength range is really challenging: not only are they invisible to the bare eye, but we still lack detectors to record them efficiently.

In the next section, I will review different single photon sources and some of their applications. I will also discuss the benefits of coupling the light source to a cavity for faster and more efficient operation. Then I will introduce some basics of photonic crystal theory. And finally I give an outlook on the rest of this work.

1.1 Single-photon sources

Single-photon emitters are light sources able to deliver triggered pulses, each consisting of exactly one photon. This unique property makes them suitable in a wide range of applications for which they surpass the qualities of other types of light sources. See references [Shields 07, Lounis 05] for a review.

1.1.1 Motivation for single-photon sources

As the light emitted by single-photon sources is amplitude-squeezed, i.e. the number of photons per pulse is well-defined, it can be used for example in the detection of weak absorption signals to reduce the shot noise on the amplitude measurement [Xiao 87, Polzik 92].

They can also be used in the generation of random numbers. Classical schemes, using pseudorandom generators or based on the noise of a physical observable, suffer from systematic errors and perturbations, that lead to deviations from a truly random distribution of numbers. The laws of quantum mechanics, however, guarantee the probabilistic collapse of a wavefunction upon measurement. For example, each individual photon incident on a 50:50 beam splitter has an equal probability to be detected in the reflected or in the transmitted path [Rarity 94, Quantis 04].

The emerging field of quantum information processing (see for example [Nielsen 00]) also benefits from the unique properties of single-photon emitters. The information is coded onto qubits (**qu**antum **bi**nary digi**ts**), a quantum mechanical state defined as a linear superposition $|\varphi\rangle = \alpha|0\rangle + \beta|1\rangle$ of the eigenstates $|0\rangle$ and $|1\rangle$ of a two level system. A measurement projects the state $|\varphi\rangle$ onto one of the basis states $|0\rangle$ or $|1\rangle$ with respective

1.1. Single-photon sources

probabilities α^2 and $\beta^2 = 1 - \alpha^2$, where $0 \leq \alpha \leq 1$. The advantage over the classical bits 0 and 1, lies in the quantum mechanical superposition, creating an infinity of possible bit states not available classically, that could be used to dramatically increase the efficiency of computing algorithms, like the factorization of large numbers for example [Ekert 96]. Another difference can be observed for two particles qubits, $|\varphi\rangle = \alpha_{00}|0\rangle_1|0\rangle_2 + \alpha_{01}|0\rangle_1|1\rangle_2 + \alpha_{10}|1\rangle_1|0\rangle_2 + \alpha_{11}|1\rangle_1|1\rangle_2$, that can form entangled states. Those are states that cannot be decomposed over the single particle basis anymore. To illustrate this, the *Bell state* $(|0\rangle_1|0\rangle_2 + |1\rangle_1|1\rangle_2)/\sqrt{2}$ cannot be obtained form the superposition of two single qubits[b]. An interesting property of this entangled state is that the measurement of the first qubit fully determines the second qubit.

Single photons are among the various candidates proposed for the fabrication of physical qubits. They have the advantages of interacting only weakly with the environment and, being propagating particles, of transporting the quantum information with them. The horizontal $|0\rangle \equiv |H\rangle$ and vertical $|1\rangle \equiv |V\rangle$ polarization of light can be used as the basis for single-photon qubits. Another possibility is to code the qubit in the spatial position of the photon, for example in one of the two output modes of a beam splitter $|0\rangle \equiv |1\rangle_T|0\rangle_R$ and $|1\rangle \equiv |0\rangle_T|1\rangle_R$.

Single photons have been used to create entangled states and demonstrate the violation of Bell's inequalities [Fattal 04b], to achieve quantum teleportation [Fattal 04a], and to realize efficient linear optic quantum computation schemes [Knill 01]. An all optical and scalable quantum CNOT gate was realized with an efficiency of 84% [O'Brien 03].

Finally, the safety in quantum key distribution [Bennett 84] for cryptography, which relies on the randomness of quantum mechanics and on the no cloning theorem [Wootters 82], also benefits from single-photon sources. Two parties that wish to communicate secretly, traditionally Alice and Bob, will encode their message by using a cipher key. To be sure that an eavesdropper, usually Eve, cannot intercept the secret key, Alice can encode it by using the polarization of *single photons*, *randomly* switching between the rectilinear ($|\uparrow\rangle$ and $|\rightarrow\rangle$) and diagonal ($|\nearrow\rangle$ and $|\searrow\rangle$) basis. Bob choses his measurement basis *randomly* and independently form Alice. Statistically, he will measure in the same basis as Alice 50% of the time. After that, both parties compare their respective basis, not the outcome of the measurement, to determine which of the transmitted bits can be used to form the

---

[b] $(a_1|0\rangle_1 + b_1|1\rangle_1) \otimes (a_2|0\rangle_2 + b_2|1\rangle_2) = a_1a_2|0\rangle_1|0\rangle_2 + a_1b_2|0\rangle_1|1\rangle_2 + b_1a_2|1\rangle_1|0\rangle_2 + b_1b_2|1\rangle_1|1\rangle_2$. As the term $|0\rangle_1|1\rangle_2$ must disappear, this means that either $a_1 = 0$ or $b_2 = 0$ and the Bell state cannot be obtained.

cypher. If Eve intercepts Alice's photons and measure them, she also has a 50% chance to get the right basis. In order to remain undetected, she needs to send new photons to Bob, introducing a 25% error rate in the line, making it possible for Alice and Bob to detect the eavesdropping. The advantage of using a single photon source is that it is intrinsically secure against photon-number-splitting attacks. However, Hwang proposed a decoy-pulse method that can be used in case of lossy communication channels or multiphoton light source to overcome this type of attack by randomly replacing signal pulses with multiphoton pulses (decoy pulses) and check for abnormal loss in those pulses [Hwang 03]. Single photon quantum cryptography experiments were first realized with diamond colour centers [Beveratos 02] and single quantum dots [Waks 02]. A comprehensive review on quantum cryptography can be found in [Gisin 02]. Commercial solutions are already available and recently, a quantum cryptography network with optical links as long as 82 km has been realized in Vienna in the frame of the integrated EU poject "SECOQC" [Anscombe 09].

A necessary device in long distance quantum communication, that can be build with single photon emitters, is the quantum repeater. They generally require small qubit circuits capable of Bell state measurements and storage of a quantum state in combination with single photon detectors and emitters. A recent scheme [Childress 06] shows that even with low efficiency, high fidelity can be reached in long-distance communication.

1.1.2 Various single-photon emitters

Attenuated lasers pulses have been used to emulate single photons sources. However, as the probability to find n photons in such a laser pulse follows the Poisson distribution, a laser source with an average number of $\langle n \rangle = 1$ photons per pulse will produce approximately one third of empty pulses, one third with one photon and one third containing two and more photons. Usually, a strongly attenuated pulse with an occupancy of $\langle n \rangle = 0.1$ photon is used to suppress ($< 10^{-2}$) the multi-photon probability. The drawback of this strategy is that around 90% of the pulses are empty, resulting in an increased level of noise and a lower efficiency.

Parametric down conversion provides an improvement as it produces correlated photon pairs [Burnham 70] through the non-linear interaction of a laser pulse with a crystal that sporadically splits a pump photon in an idler and a signal photon. The first one can be used to indicate the presence of his companion, for example by triggering a detector only when the pulse is not empty, thus reducing the dark noise [Hong 86].

1.1. Single-photon sources

However, as for the attenuated lasers, care should be taken not to produce multiple pairs per pulse, which limits the production rate.

Atoms, ions, and some organic molecules have discrete energy levels, i.e. there is no allowed energy state between them. When an electron decays from an excited energy level to a lower energy level, it needs to release the excess energy in one shot. This energy quantum is nothing else than a photon [Einstein 05], whose frequency ν is related to the energy gap through $E_{\text{gap}} = h\nu$.

Actually, the first single-photons were measured on a cascade transition in calcium atoms [Clauser 74]. Anti-bunching was then observed on an attenuated sodium atom beam [Kimble 77]. An improvement came with the use of single ion traps, allowing long observation times on the same ion. However, these systems are not practical for integration.

Some organic dyes emit fluorescence between the LUMO and HOMO with a high quantum yield and show very strong anti-bunching effect both at low and at room temperatures. At room temperature however, even for molecules protected from oxygen and held in a polymer matrix, the stability is in the order of hours [Lounis 00b].

Colour centers are defects of insulating inorganic crystals with intense absorption and fluorescence bands or lines, very similar in structure as for organics molecules. Nitrogen-vacancies (NV) centers and nickel-nitrogen complex (NE8) in diamond have been proposed as single-photon emitters. They offer long stability and high quantum yields, however, the spontaneous emission lifetime is rather long (< 10 ns) and they have a dark state responsible for bunching in the correlation function and limiting the fluorescence intensity. Their emission in the visible is not suitable for fiber based communication and the high refractive index of diamond impairs the extraction of the photons.

Nanocrystals of II-VI materials, such as CdSe-ZnS, are colloidal crystalline semiconductor structures, a few nanometers in diameter, containing thousands of atoms [Klimov 04]. They have a size-dependent narrow emission line spectrum and a broad absorption continuum above the exciton transition. Due to the high efficiency of Auger processes, they do not exhibit multiple exciton emission. They show anti-bunching emission [Michler 00a, Lounis 00a] at room temperature. They are easy to synthesize, however, relatively long lifetime, blinking and spectral hopping [Shimizu 01] limit their application as single photon emitters.

Single walled carbon nanotubes were recently shown to posses photoluminescence anti-bunching at low temperature[Högele 08] and short emission lifetime (10-20 ps). However, like for nanocrystals, they suffer from intensity and spectral fluctuations.

Self-assembled quantum dots (QDs) are 3-dimensional confined semiconductor structures embedded in a semiconductor crystal with a higher energy gap, usually InAs in GaAs with emission around 900 nm. A major advantage of these structures over the other light sources presented above, is the possibility of integration and electrical contacting readily available in III/V technology. Anti-bunching was first observed under optical pumping [Michler 00b, Santori 01, Thompson 01], soon followed by electrical injection [Yuan 02]. The experiments were performed at liquid helium temperatures to avoid emission broadening from the interaction with phonons. The low temperatures might be a problem in some applications. However, single-photon generation has been realized for temperatures in excess of 100K[Mirin 04] and recently, [Dou 08] demonstrated single-photon operation on a LED at 77 K.

1.1.3 Self-assembled QDs with emission around 1.3 μm

In this work, we use self-assembled InAs QDs with emission wavelength around 1300 nm. This wavelength range, known as the second telecommunication window, corresponds to very low optical losses and minimal chromatic dispersion in optical fibers.

The QDs are grown by molecular beam epitaxy in a GaAs matrix: In and As adatoms are added under ultra-high vacuum on the crystalline surface of the GaAs substrate. Since these materials have different lattice parameters, strain will build up in the crystal, leading to the formation of small InAs islands randomly positioned over a thin wetting layer. This is known as the Stranski–Krastanov growth mode[Stranski 39, Pimpinelli 98]. The sample is then overgrown with GaAs.

Since InAs has lower energy band edges than GaAs, these small islands, the Quantum Dots (QDs), can confine the charge carriers. The 3-dimensional confinement at nanoscale dimensions gives atomic like properties to the QDs, as it leads to the creation of quantized energy levels for electrons and holes. The Coulomb interactions between electron and hole pairs further binds them as excitons. The wavelength of the photon emitted as the electron-hole pair recombines is thus dependent on the size of the QD, usually in the 850–1000 nm

1.1. Single-photon sources

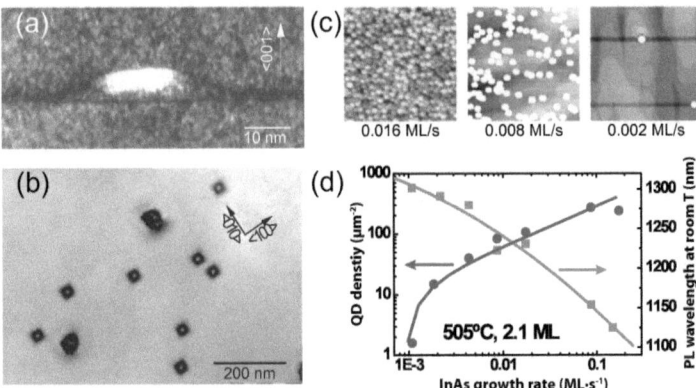

Figure 1.1: (a) Cross-section and (b) plan-view TEM dark-field images of a QDs grown at low InAs growth rate (0.0015 ML/s) and capped by GaAs [Alloing 07]. (c) $1 \times 1\,\mu m^2$ AFM images of 2.1 ML InAs at different deposition rates. (d) QD density, as measured from AFM images, and evolution of PL emission wavelength as a function of InAs growth rate.[Alloing 05]

range [Bayer 02a].

However, as depicted on figure 1.1, by lowering the growth rate it is possible reduce the QD density to a few QDs per square micrometers while increasing their size [Alloing 05]. This is noticeable as a wavelength shift of the ground state emission of the QDs to the 1300–1400 nm range: in larger dots, the confinement decreases, leading to a lowering in the energy of the quantized levels. TEM measurements (see figure 1.1a and b) show that the QDs have a truncated pyramidal shape, with a basis length of 12–17 nm and a height around 7.5–9 nm.

A big difference, as compared to nanocrystals where Auger processes are very efficient, is that the QDs can sustain multiple excitons. Due to the Pauli principle, two electrons with opposite spins are allowed per energy level. The energy degeneracy is lifted by the Coulomb interaction between the exciton and additional charges. At low temperature, this results in a photoluminescence spectrum with discrete narrow (tens of μeV) emission lines corresponding to the recombination of the exciton (X), the biexciton (XX) and charged excitons (X^+, etc.) as seen on figure 1.2.

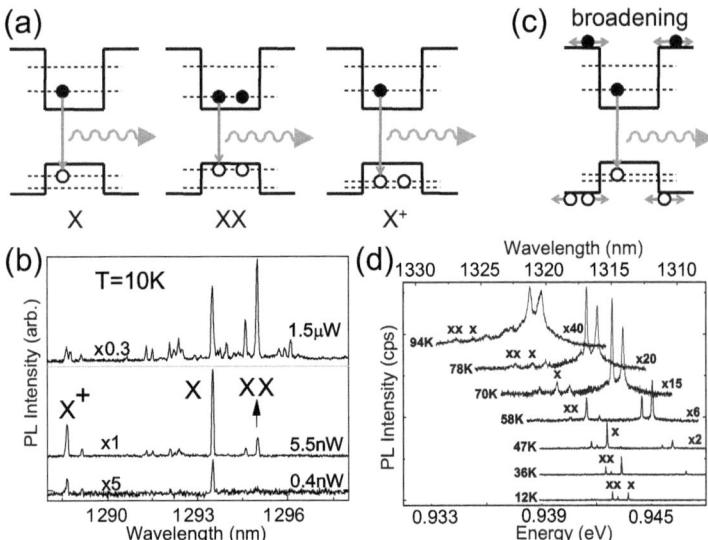

Figure 1.2: (a) Example of atom-like transitions in a QD.(b) Photoluminescence (PL) spectrum of a single QD at various excitation power [Zinoni 06]. Notice the background appearing at high excitation. (c) Broadening mechanism due to Coulomb interaction with fluctuating environmental charges [Kamada 08]. (d) Temperature dependent broadening due to interactions with phonons [Alloing 05].

To prove that these transitions emit single photons, we can measure the second-order correlation function $g^{(2)}(\tau)$ in a Hanbury-Brown and Twiss interferometer [Loudon 00]: the photons are sent to a beam splitter, with a single-photon detector at each output, and the delay between detection events is recorded. In the case of a single-photon source, a dip to 0 should appear at zero time delay[c], since it is not possible to record a single photon on both detectors simultaneously. It is thus important to make sure that only one excitonic line is selected. This is usually done by spectral filtering. The X, XX, and X$^+$ each show separately a clear anti-bunching dip at zero delay (see figure 1.3).

---

[c]$g^{(2)}(0) = 1 - 1/n$ for a n-photon source, and $g^{(2)}(\tau) \geq 1$ for classical light sources. [Loudon 00]

1.1. Single-photon sources

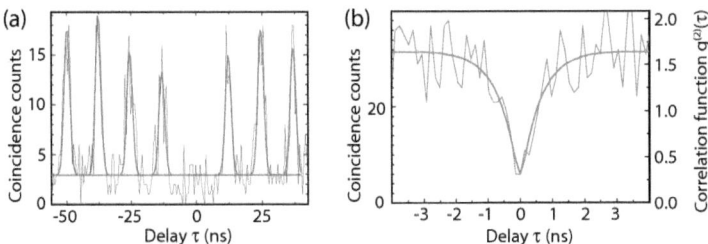

Figure 1.3: (a) Anti-bunching behavior of an exciton (X) under pulsed excitation, and (b) of a charged exciton (X^+) under continuous excitation. [Zinoni 07]

Self-assembled QDs, though having a quantum efficiency close to unity, suffer from low photon extraction due to the high refractive index contrast between GaAs and air (see figure 1.4). A simple calculation shows that only approximately 2% of the emitted photons will leave the top surface of the sample, because of total internal reflection at the interface. From those, only about 13% can be collected by a microscope objective with a numerical aperture of NA=0.5. This is in the best case, neglecting reflections and losses from the objective. This means that about 1 pulse out of 1000 contains a single photon, the others being empty. Adding the emission lifetime around 1 ns, this limits the collection rate below the MHz range.

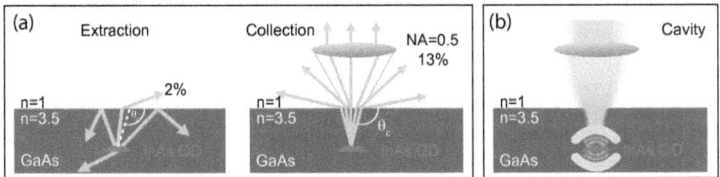

Figure 1.4: (a) If we neglect reflections at interfaces, approximately 2% of the photons emitted by the QD escape the GaAs and 13% of them can then be collected in an optic with a numerical aperture of 0.5. (b) We can couple the QD to a microcavity. Coupled photons will be funneled out of the GaAs through the cavity mode. The far field pattern can be engineered for efficient collection.

1.2 Purcell effect

A possible path to solve this problem is to embed the QDs into a microcavity [Andreani 99, Gérard 01, Vahala 03, Khitrova 06]. The content of this section is mainly inspired by [Benisty 98].

The equations describing the coupling of an emitter and a cavity are identical to the evolution equations of two damped classical oscillators with the same oscillation frequency, which are linearly coupled to each other. They can have two different regimes, depending on their damping constants and their coupling. In the strong coupling regime, the system shows two different oscillation frequencies, but is damped with a fixed average rate. In the weak coupling regime, the system oscillates at a fixed frequency, but with different decay rates.

For the cavity-emitter system, if the cavity were perfect (strong coupling regime), the system would experience a Rabi oscillation at the angular frequency Ω, periodically exchanging the photon energy between the cavity mode and the emitter. On resonance, their spectrum becomes a doublet with equal linewidths. The need for high quality cavities makes strong coupling difficult to observe in solid state physics, but was nonetheless demonstrated recently in micropillars [Reithmaier 04], in photonic crystal cavities [Yoshie 04, Hennessy 07, Englund 07, Winger 08], and in microdiscs [Peter 05, Srinivasan 07].

In the "bad" cavity limit (weak coupling regime), i.e. when Ω is much smaller than the cavity linewidth, the evolution of the emitter to its ground state is exponential but occurs at a different rate as in free space. The cavity modifies the electromagnetic environment of the QD. The density of optical states available for the QD to emit a photon is increased at some resonance frequencies as compared to the vacuum. On resonance, the QD has more possibilities to emit a photon in the cavity mode than in vacuum. Its spontaneous emission rate will increase. On the other hand, when out of resonance, the optical modes are be sparser. The lifetime of the emitter gets longer, as the possibility to emit a photon is reduced. The Purcell effect, after the name of its discoverer [Purcell 46], has already been widely observed with self-assembled QD with emission below 1050 nm in different types of cavities. See for example [Gérard 98, Gérard 99, Graham 99, Gayral 01, Kiraz 01, Bayer 01, Solomon 01, Englund 05, Kress 05, Gevaux 06, Chang 06].

The mode has also a spatial extension: some regions have a higher field than others. Even if the QD has the right energy, it also need to be located at a maximum of the mode to obtain large coupling. The random nucleation sites of self-assembled QDs makes the

1.2. Purcell effect

control of the spatial coupling really difficult. One strategy is to measure the position of the QD and build the cavity around it [Badolato 05]. We chose the statistical approach, consisting in defining a large number of cavities per sample and measuring them all.

When a photon is emitted in the cavity mode, it will then be efficiently funneled out of the GaAs crystal through the mode radiation pattern. In addition, the far-field of the mode can be engineered with a small emission angle to match the numerical aperture of the collection optics.

1.2.1 The Purcell factor

The radiative lifetime of a dipole emitter transition between an initial $|i\rangle$ and a final $|f\rangle$ state can be estimated with the Fermi Golden Rule,

$$\frac{1}{\tau} = \frac{2\pi}{\hbar^2} \left|\langle f|\hat{H}|i\rangle\right|^2 \rho(\omega_e) \tag{1.1}$$

where $\rho(\omega_e)$ is the density of optical modes[d] at the emitter angular frequency ω_e, and $\left|\langle f|\hat{H}|i\rangle\right|$ the transition matrix element. This approach is valid when the emitter "sees" a quasi-continuum of modes, i.e. when its linewidth is narrower that the linewidth of the cavity mode.

The cavity perturbs the density of optical modes both energetically and spatially. They are concentrated at certain resonant frequencies and locations, where we thus expect an increase of the spontaneous emission rate over free space. On the other hand, the rate is expected to be lower where the optical modes are sparser. An example of a calculated density of states can be found on figure 1.10.

We can evaluate equation (1.1) to estimate the decay rate of the emitter. The Jaynes-Cummings atom-field hamiltonian can be decomposed as $\hat{H} = \hat{H}_{\text{atom}} + \hat{H}_{\text{field}} + \hat{H}_{\text{int}}$, and we may consider[e] only the interaction $\hat{H}_{\text{int}} = -\hat{\mathbf{E}}(\mathbf{r}_e) \cdot \hat{\mathbf{D}}$ between the atom at position $\mathbf{r}_e$ and the field. $\hat{\mathbf{D}}$ is the dipole atomic operator. The electric field in the mode is

$$\hat{\mathbf{E}}(\mathbf{r}) = \mathcal{E}_{max} \left[\hat{a}\mathbf{f}^*(\mathbf{r}) + \hat{a}^\dagger \mathbf{f}(\mathbf{r})\right], \tag{1.2}$$

where $\hat{a}$ and $\hat{a}^\dagger$ are the photon annihilation and creation operators and $\mathbf{f}(\mathbf{r})$ is a dimensionless complex vector function which describes the mode spatially. Its modulus is normalized to

---
[d] $\rho(\omega_e) = \frac{dn}{d\omega_e} = \hbar\frac{dn}{dE_e} = \hbar\rho(E_e)$
[e] rotating wave and electric dipole approximation

unity at the field maximum.

$$\mathcal{E}_{max} = \left(\frac{\hbar\omega}{2\varepsilon_0 \mathcal{V}}\right)^{\frac{1}{2}}, \text{ with } \mathcal{V} = \int_\mathbf{r} n(\mathbf{r})^2 |\mathbf{f}(\mathbf{r})|^2 \, d^3\mathbf{r} \qquad (1.3)$$

is derived by comparing the energy of the radiation field before and after quantization [Loudon 00], and $\mathcal{V}$, the effective mode volume, describes the concentration of the electrical field in a restricted space with refractive index $n(\mathbf{r})$.

If we examine spontaneous emission, i.e. the relaxation of the emitter from the excited state $|e\rangle$ to the ground state $|g\rangle$ by emitting one photon, considering (1.2) and the dipole atomic operator

$$\hat{\mathbf{D}} = \mathbf{d}\left(|e\rangle\langle g| + |g\rangle\langle e|\right), \qquad (1.4)$$

where $\mathbf{d} = \langle e|\mathbf{q\hat{r}}|g\rangle = \langle g|\mathbf{q\hat{r}}|e\rangle$ is the matrix element of the dipole operator, we can evaluate the matrix element

$$\left|\langle f|\hat{H}|i\rangle\right| = \left|\langle 1, g| - \hat{\mathbf{E}} \cdot \hat{\mathbf{D}}|0, e\rangle\right| = \mathcal{E}_{max} |\mathbf{d} \cdot \mathbf{f}^*(\mathbf{r}_e)| = \hbar\Omega. \qquad (1.5)$$

Ω is the Rabi frequency mentioned at the beginning of this section.

In free space, the density $\rho(k)\mathrm{d}k$ of field mode, defined as the number of modes per unit volume $\mathcal{V}$ having their wavevector in the specified range, is obtained by introducing an arbitrary cavity $\mathcal{V}$ which discretizes the solutions of the wave equation. By counting the number of solutions with wavevector values between k and $k + \mathrm{d}k$ and taking into account the two polarizations, we find $\rho(k)\mathrm{d}k = k^2\mathrm{d}k/\pi^2$ [Loudon 00]. As the angular frequency is given by $\omega = c_0 k/n$, we obtain $\rho(\omega)\mathrm{d}\omega = (\omega n/\pi c_0)^2 \cdot (n/c_0)\mathrm{d}\omega$. The mode density in an homogenous isotropic material (i.e. $|\mathbf{f}(\mathbf{r})| = 1$), can thus be written as

$$\rho_0(\omega) = \frac{\omega^2 \mathcal{V} n^3}{\pi^2 c_0^3} \qquad (1.6)$$

and the decay rate is then

$$\frac{1}{\tau_0} = \frac{2\pi}{\hbar} \cdot \frac{\omega^2 n^3}{\pi^2 c_0^3} \cdot \frac{\hbar\omega}{2\varepsilon_0} |\mathbf{d}|^2 \cdot \frac{1}{3} \qquad (1.7)$$

where the final factor $1/3$ represents the random orientation of the modes within a uniform dielectric with respect to the dipole.

In the case of a single-mode cavity with a quality factor Q, defined as $Q = \omega_c/\Delta\omega_c$, the mode density seen by a narrow emitter is given by the normalized Lorentz function

$$\rho_{cav}(\omega) = \frac{2Q}{\pi\omega_c} \cdot \frac{\Delta\omega_c^2}{4(\omega - \omega_c)^2 + \Delta\omega_c^2} \qquad (1.8)$$

1.2. Purcell effect

and the spontaneous emission rate is

$$\frac{1}{\tau_{cav}} = \frac{2\pi}{\hbar} \cdot \frac{2Q}{\pi\omega_c} \cdot \frac{\Delta\omega_c^2}{4(\omega-\omega_c)^2 + \Delta\omega_c^2} \cdot \frac{1}{V_{cav}} \cdot \frac{\hbar\omega}{2\varepsilon_0} \cdot |\mathbf{d} \cdot \mathbf{f}(\mathbf{r}_e)|^2 \qquad (1.9)$$

Comparing expressions (1.7) and (1.9), we see that the emitter's spontaneous emission rate is enhanced (or inhibited) by a factor:

$$\frac{\tau_0}{\tau_{cav}} = \underbrace{\frac{3Q(\lambda_c/n)^3}{4\pi^2 V_{cav}}}_{F_p} \cdot \underbrace{\frac{\Delta\omega_c^2}{4(\omega-\omega_c)^2 + \Delta\omega_c^2}}_{\leq 1} \cdot \underbrace{\xi^2 |\mathbf{f}(\mathbf{r}_e)|^2}_{\leq 1} \qquad (1.10)$$

where $\xi = \frac{|\mathbf{d}\cdot\mathbf{f}(\mathbf{r}_e)|}{|\mathbf{d}||\mathbf{f}(\mathbf{r}_e)|}$ describes the orientation matching of $\mathbf{d}$ and $\mathbf{f}(\mathbf{r}_e)$. The second term of (1.10) accounts for the frequency detuning and the third one describes the orientation matching between the dipole and the field at the location of the emitter. This third term also accounts for the spatial coupling, since it indicates that the effect is at the highest when the emitter is at a maximum of the field.

The first term, known as the Purcell factor

$$F_p = \frac{3Q(\lambda_c/n)^3}{4\pi^2 V_{cav}}, \qquad (1.11)$$

is related only to the physical properties of the cavity and gives the maximum rate enhancement expected in the case of a dipole perfectly aligned to the cavity mode polarization and energy.

1.2.2 Corrections for Quantum Dots

If we now consider an emitter at the anti-node of the field, or with a large frequency mismatch, equation 1.10 predicts the complete suppression of its spontaneous emission. However, real cavities are not perfect and they can support leaky modes. The Fermi Golden Rule (1.1) for the cavity can be modified accordingly [Imamoğlu 99]:

$$\frac{1}{\tau_{cav}} = \frac{2\pi}{\hbar^2} \mathcal{E}_{max} |\mathbf{d} \cdot \mathbf{f}^*(\mathbf{r}_e)|^2 \rho_{cav}(\omega_e) + \frac{1}{\tau_{leak}} \qquad (1.12)$$

Obviously, the lifetime of the detuned emitter allows a direct estimation of τ_{leak}, since in this case $(1/\tau_{cav}) = 0 + (1/\tau_{leak})$.

Self-assembled QDs are not isotropic and, due to the vertical confinement, the transition dipole element $\mathbf{d} = (d_x, d_y, 0)$ is aligned perpendicularly to the growth axis[f]. The random in plane polarization of the dipole gives $\xi^2 \simeq 1/2$.

---
[f] $d_z = \langle e|\hat{z}|hh\rangle = 0$ for transitions to the heavy holes.

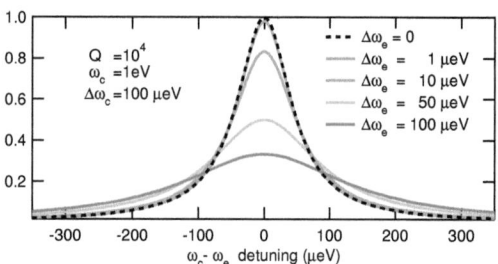

Figure 1.5: Effect of the emitter linewidth on the cavity coupling efficiency. The black dotted line correspond to an ideal emitter (equation (1.13) left)

As mentioned in section 1.1.3, the linewidth of the excitonic emission homogeneously broadens with the temperature through interaction with the phonons (figure 1.2d). But we also observed an excitation intensity dependent broadening (figure 1.2b and 5.4b). This broad "background" feeds the cavity mode when it is off resonance. Hennessy et al. reported anti-bunching between cavity and exciton emission, confirming the strong correlation between the exciton and the background [Hennessy 07]. It most probably originates from interactions between the exciton and randomly fluctuating charges around the QD (figure 1.2c). A correction to the cavity density of states to account for the broadening of the emitter has been proposed [Ryu 03] (see figure 1.5):

$$\frac{\Delta\omega_c^2}{4(\omega-\omega_c)^2+\Delta\omega_c^2} \xrightarrow{\text{broad emitter}} \frac{1+2Q\gamma_e}{4Q^2(1-\omega_e/\omega_c)^2+(1+2Q\gamma_e)^2} \quad (1.13)$$

where $Q = \omega_c/\Delta\omega_c$ is the quality factor of the cavity and $\gamma_e = \Delta\omega_e/\omega_e$ is directly proportional to the linewidth of the emitter.

In the case of a QD, we can finally rewrite the averaged spontaneous emission ratio as:

$$\frac{\tau_0}{\tau_{cav}} = \frac{1}{2} \cdot F_p \cdot \frac{1+2Q\gamma_e}{4Q^2(1-\omega_e/\omega_c)^2+(1+2Q\gamma_e)^2} \cdot |\mathbf{f}(\mathbf{r}_e)|^2 + \frac{\tau_0}{\tau_{leak}} \quad (1.14)$$

with F_p the Purcell factor introduced in (1.11).

If a cavity mode is polarization degenerate, this provides an additional channel for photon emission, and the rate of the coupled emitter is thus increased accordingly.

1.2.3 Coupling efficiency

The mode coupling efficiency, i.e. the ratio of emitted photons coupled to the mode, can be estimated as

$$\beta = \frac{\tau_{cav}^{-1}}{\tau_{cav}^{-1} + \tau_{leak}^{-1}} \simeq 1 - \frac{\tau_{cav}}{\tau_{leak}} \qquad (1.15)$$

by measuring τ_{cav} when a QD is on resonance with the mode and τ_{leak} when its frequency is detuned.

1.3 Photonic Crystals

A photonic crystal is a periodic variation of the refractive index $n(\mathbf{r}) = \sqrt{\epsilon(\mathbf{r})}$ in 1, 2, or 3 dimensions [Joannopoulos 95]. They can be used to modify the flow of photons through a material, as light waves are sensitive to n.

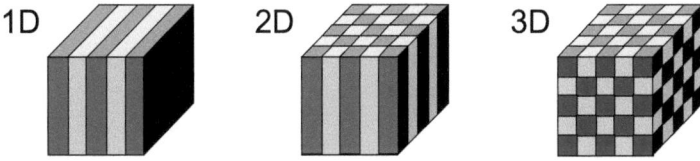

Figure 1.6: Illustrations of photonic crystals in 1, 2 and 3 dimensions realized by periodically varying the refractive index n.

1.3.1 The master equation

The propagation of light in matter is governed by the four Maxwell equations

$$\begin{aligned} \nabla \cdot \mathbf{B} &= 0 & \nabla \times \mathbf{E} + \frac{1}{c}\frac{\partial \mathbf{B}}{\partial t} &= 0 \\ \nabla \cdot \mathbf{D} &= 4\pi\rho & \nabla \times \mathbf{H} - \frac{1}{c}\frac{\partial \mathbf{D}}{\partial t} &= \frac{4\pi}{c}\mathbf{J} \end{aligned} \qquad (1.16)$$

where $\mathbf{B}$ and $\mathbf{D}$ are the magnetic induction and displacement fields, $\mathbf{E}$ and $\mathbf{H}$ are the macroscopic electric and magnetic fields, and ρ and $\mathbf{J}$ are the free charges and currents.

For photonic crystals, we can consider a low-loss mixed dielectric medium, i.e. a composite of homogeneous regions with no free charges or currents, for which the $\mathbf{D}(\mathbf{r}) = \varepsilon(\mathbf{r})\mathbf{E}(\mathbf{r})$ and $\mathbf{B} = \mathbf{H}$. This is the case for most semiconductors, including (Al)GaAs and InA. Their refractive index n is the square root of scalar dielectric constant $\varepsilon(\mathbf{r}, \omega)$.

The Maxwell equations (1.16) can then be simplified to

$$\nabla \cdot \mathbf{H}(\mathbf{r}) = 0 = \nabla \cdot \mathbf{D}(\mathbf{r}) \quad (1.17)$$

and the *master equation*

$$\nabla \times \left(\frac{1}{\varepsilon(\mathbf{r})}\nabla \times \mathbf{H}(\mathbf{r})\right) = \left(\frac{\omega}{c}\right)^2 \mathbf{H}(\mathbf{r}) \quad (1.18)$$

Equation (1.17) requires that the field configuration is built up of transverse electromagnetic waves, i.e. with $\mathbf{E}$ and $\mathbf{B}$ perpendicular to the direction of propagation $\mathbf{k}$, and equation (1.18) can be seen as an eigenvalue problem $\hat{\Theta}\mathbf{H}(\mathbf{r}) = (\omega/c)^2 \mathbf{H}(\mathbf{r})$ with the hermitian operator

$$\hat{\Theta} = \nabla \times \left(\frac{1}{\varepsilon(\mathbf{r})}\nabla \times \right) \quad (1.19)$$

This is particularly interesting, as it guarantees that the solutions are orthogonal and with real eigenvalues $(\omega/c)^2$. We also notice that the there is no fundamental length scale contained in (1.18), which makes the problem fully scalable (see section 3.1).

It is possible to derive a master equation for the electric field as well. However, the operator is not hermitian in this case. Still, the electric field can be retrieved from the solutions of (1.18) through $\mathbf{E}(\mathbf{r}) = (-ic/\omega\varepsilon(\mathbf{r})) \nabla \times \mathbf{H}(\mathbf{r})$.

By using the Bloch theorem, the periodicity $\varepsilon(\mathbf{r}+\mathbf{R}) = \varepsilon(\mathbf{r})$ of the photonic crystal can be translated to the solutions of (1.18) as

$$\mathbf{H}_k(\mathbf{r}) = \mathbf{u}_k(\mathbf{r})e^{i\mathbf{k}\cdot\mathbf{r}} \quad (1.20)$$

where $\mathbf{u}_k(\mathbf{r})$ has the same periodicity and symmetries as $\varepsilon(\mathbf{r})$.

The periodicity of $\varepsilon(\mathbf{r})$ and the hermitian properties of (1.18) remind strongly of the quantum mechanical treatment of the free electron problem in a crystal. Indeed, under certain conditions, we will see the formation of a forbidden energy band for photons, pretty much as the bandgap for electrons in semiconductors.

As the vector operator $\hat{\Theta}$ is not separable, numerical approaches are usually required to solve this problem. The most common is the finite-difference time-domain (FDTD) algorithm which computes the discretized Maxwell equations in time and space on a mesh representing the photonic crystal. Figure 1.10 shows the result of calculations performed with a 3D Finite Element Maxwell (FEM) solver [Römer 07b], which, unlike the FDTD method, does not suffer from the discretization in time. Another approach is the plane-wave-expansion method, which uses the periodicity of the crystal to solve the problem in the reciprocal space after decomposing $\mathbf{H}(\mathbf{r})$ and $\varepsilon(\mathbf{r})$ in their Fourier components. The band diagrams and the majority of the mode-field patterns presented in this work were solved with a PWE routine written by V. Zabelin[g].

1.3.2 Two-dimensional photonic crystals

We will illustrate the case of a two-dimensional triangular photonic crystal of air holes in a dielectric media, as it is particularly relevant to this thesis. A schematic representation of such a crystal with lattice parameter a is provided on figure 1.7. The holes extend indefinitely in the z direction, perpendicular to the crystal plane.

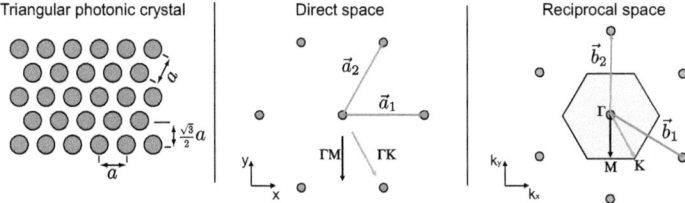

Figure 1.7: Two-dimensional triangular photonic crystal with circular holes. The lattice parameter is a. The reciprocal lattice $(\vec{b}_1, \vec{b}_2)$ can be constructed form the direct basis vectors $(\vec{a}_1, \vec{a}_2)$. The first Brillouin zone has hexagonal symmetry and with high-symmetry points Γ, M, and K.

Similarly to solid state physics [Kittel 04], we can construct the reciprocal space in the usual way. The symmetry of the crystal allows us to examine only the region delimited by the high symmetry points Γ, M, and K, with $|\Gamma M| = \frac{2}{\sqrt{3}}\frac{\pi}{a}$ and $|\Gamma K| = \frac{4}{3}\frac{\pi}{a}$, as defined in figure 1.7. In a homogeneous medium, the light wave propagating along the direction $\mathbf{k}$ has a frequency $\omega = ck$, directly proportional to $k = |\mathbf{k}|$. In a photonic crystal however, the relation is not as simple. Figure 1.8 shows this dispersion relation between $\mathbf{k}$ and ω for a

---
[g]Laboratory of Quantum Optoelectronic, EPFL

photonic crystal of air holes in a material with refractive index $n = 3.41$, corresponding to bulk GaAs for light around 1300 nm.

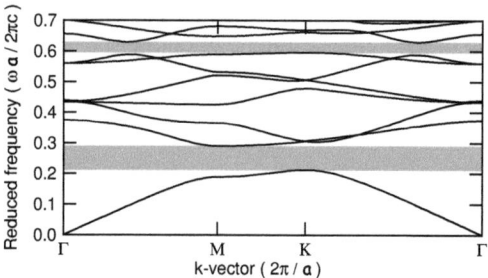

Figure 1.8: Photonic band structure for a triangular lattice of holes in a material of refractive index $n = 3.41$. No wavevector **k** matches the frequencies of the photonic bandgap (red shaded zones).

We observe the formation of photonic band gaps: light in this frequency range cannot propagate through the photonic crystal. This 2-dimensional generalization of a Bragg-mirror can thus reflect the light independently of the incidence angle.

We use this property to build a cavity, for example by omitting several holes in the middle of the structure. A photon within the energy gap could not leave the cavity in radial direction, and would travel only perpendicularly to the air holes. This is the principle of a photonic crystal fiber.

1.3.3 Two-dimensional photonic crystals in membrane

We can generalize these results and build a 3-dimensional photonic crystal to trap light in all dimensions. However, this is difficult to achieve within the III/V semiconductor technology. Still we can find other strategies to confine the light. For example, we can drill the holes into a planar waveguide. The vertical confinement will then be provided by the propagating modes of the waveguide.

As this configuration is different from an ideal photonic crystal, we need to apply some corrections to our model. We can separate the problem and solve first the waveguiding conditions to obtain an effective refractive index n_{eff} for the propagating mode. We then

1.3. Photonic Crystals

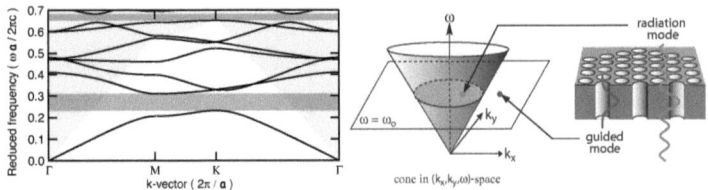

Figure 1.9: (left) Photonic band structure for a triangular lattice of holes in a material of refractive index $n = 3.11$, corresponding to the effective index of the fundamental guided mode in a 330 nm thick GaAs waveguide. The shaded area corresponds to the light cone. (right) Illustration [Srinivasan 02] of guided and radiation modes of a membrane waveguide, together with their position with respect to the light cone.

compute the 2-dimensional photonic crystal structure again, using n_{eff} instead of the refractive index of GaAs. Figure 1.9 shows that the new band structure has narrower bandgaps. Intuitively, the smaller the refractive index difference between the dielectric and the holes, the narrower the bandgap will be. Variations in the relative size of the hole has a similar effect on the bandgap. See for example figure 3.2 on page 50.

Membranes do not confine the light as efficiently as a photonic crystal, and they can support some radiation modes, as depicted on the right part of figure 1.9. The frequency range above the light cone can be reached for large vertical component k_z of the wavevector **k**. As $|\mathbf{k}| = n \cdot \omega/c$, this means that for a fixed coordinate (k_x, k_y), k_z is proportional to ω. When $k_z > \omega_0$, the photons propagate above the critical angle θ_c defined on figure 1.4 and can escape to free space.

1.3.4 Photonic crystal cavities

Introducing an impurity in a semiconductor provokes the creation of donor and acceptor states in the bandgap. Similarly, a defect inserted in the photonic crystal, for example by not removing three holes in the ΓK direction as for so-called L3 cavities, creates modes with frequencies within the photonic gap.

Figure 1.10a shows the spatial distribution of the local density density of optical states (LDOS) at a frequency corresponding to a mode of the L3 cavity. The arrows indicate the preferred polarization direction. We notice that the field is concentrated in the defect region

30 Chapter 1. Introduction

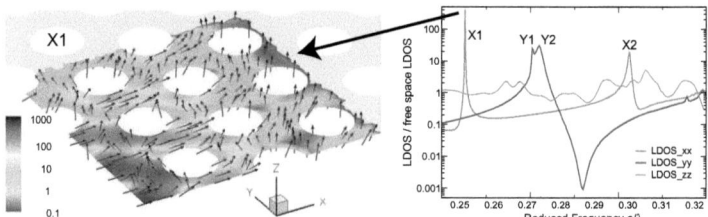

Figure 1.10: (a) Spatial distribution of the local density of states (LDOS) corresponding to the lowest resonance of L3 cavity in a waveguide. The color scale is in logarithmic units and the arrows indicate the preferred polarization. (b) LDOS ratio at the center ($x = y = 0$) of the cavity for frequencies in the photonic bandgap [Römer 07b].

and does not extend far in the "two-dimensional mirror" formed by the photonic crystal. We see also that this particular mode is mainly polarized in the x direction.

If we look at the center of the cavity for different reduced frequencies (figure 1.10b), we notice strong variations in the relative LDOS, corresponding to 4 different modes sustained by the L3 defect. The vertically polarized LDOS (green curve) is not sensibly different from that of free space: as expected, the confinement provided by the waveguide is not very strong. However, this should not influence the rate enhancement (equation 1.10), as the QDs dipole moment is in plane.

The cavity may suffer from losses due to coupled photons escaping to the light cone. However, by carefully engineering the shape of the cavity, we can reduce the amount of losses. For example [Akahane 03] proposed to modify the length of a L3 cavity to obtain a better confinement of the mode. This can be best seen in the reciprocal space (figure 1.11). For the modified L3 design, the electric field of the mode has almost no **k** values corresponding to the light cone (leaky region), as compared to the unmodified case.

Photonic crystal is a versatile tool in building cavities, as it allows for a large flexibility in their design (see section 3.1). In chapter 5, we will also demonstrate QD to mode coupling by measuring the Purcell rate enhancement. In chapter 4, we will see that by tuning the shape of a cavity, it is also possible to tailor the far-field of the mode, and thus collect the coupled photons efficiently.

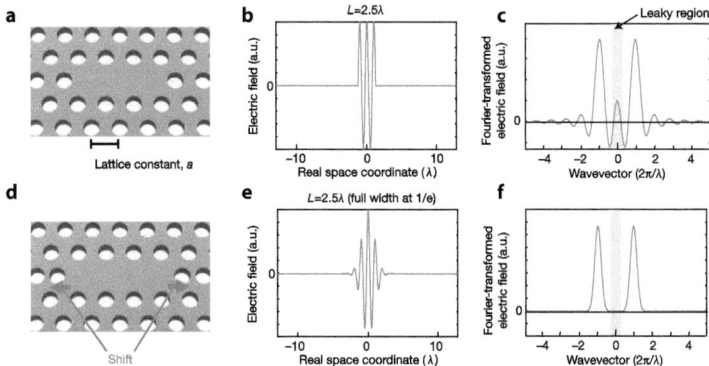

Figure 1.11: (a) Illustration of a L3 cavity formed by not removing 3 holes in the ΓK direction. (b) The electric field along ΓK is strongly confined by the borders of the cavity which results (c) in a non-vanishing amplitude of its Fourier-transform in the light cone. (d) A modified cavity has (e) a smoother electric field profile. (f) The coupling to the light cone is strongly reduced. (from [Akahane 03])

1.4 Outline of the manuscript

Rate enhancement of the photoluminescence of InAs quantum dots due to the Purcell effect had already been demonstrated in various cavity types (see table 1.1), starting with Gérard et al. in 1998. InAs QDs were used as an internal light source to probe the modes of photonic crystal cavities since 2001. And, coinciding with the beginning of my thesis work, strong coupling was demonstrated in pillar and in photonic crystal microcavities.

In the following pages, I will present our efforts in coupling single QDs with emission at 1 300 nm in photonic crystal cavities. Next chapter describes the fabrication of photonic crystal on GaAS membranes and our measurement setups. The third chapter is dedicated to different tuning strategies to bring the emitter and cavity to resonance. The fourth chapter will focus on our efforts to increase the collection efficiency of the photons by modifying the structure of the photonic crystal cavity. Finally, and before concluding, the fifth chapter concerns our experimental results on the emission rate enhancement of cavity coupled QDs under optical and electrical operation.

The first part of this project was realized at the *Ecole Polytechique Fédérale de Lausanne* (EPFL, Switzerland) and was continued at the *Technical University of Eindhoven* (TU/e, Netherlands) after our group moved there. This work profited from the collaboration with:

A. Gerardino and M. Francardi *Institute of Photonics and Nanotechnologies*, CNR, Roma, Italy, for e-beam lithography of the photonic crystals and for the LED processing.

F. Römer *Integrated system laboratory*, ETH Zurich, Switzerland, for simulations on the 3D Finite Element Maxwell solver.

S. Vignolini and F. Intonti *LENS and Department of Physics*, University of Florence, Italy, for the SNOM studies of our photonic crystal micro-cavities.

P. El-Kallassi *Laboratoire d'Optoelectronique des Matériaux Moléculaires*, EPF Lausanne, Switzerland, for polymer infiltration experiments on or structures

A. Rastelli *Max-Plack-Institute für Festkörperforschung*, Stuttgart, Germany, for in-situ thermal annealing experiments on our cavities.

Finally, I performed micro-photoluminescence (micro-PL) and time-resolved measurements together with **N. Chauvin**. The quantum dot structures were grown in our group by **B. Alloing** and **L. H. Li**. I could also profit from the micro-PL setup built by **C. Zinoni**, and from **D. Bitauld**'s cryogenic probe-station. **C. Monat** introduced me to photonic crystals and assisted me in the realization of the first mask.

1.4. Outline of the manuscript

Table 1.1: Coupling of InAs quantum dots to various cavity types. The measured rate enhancement is indicated, as well as the emission wavelength. ab indicates anti-bunching experiments and $str.\ cpl.$ the strong coupling regime.

Cavity type	τ_0/τ_{cav}	Wavelength	Reference
pillar microcavity	5×	930 nm	[Gérard 98]
oxyde-apertured microcavity	2.3×	990 nm	[Graham 99]
pillar microcavity	4.6×	896 nm	[Solomon 01]
photonic crystal cavity	–	950 nm	[Reese 01]
microdisk	12×	996 nm	[Gayral 01]
microdisk	6×	938 nm	[Kiraz 01]
photonic crystal cavity	–	1 200 nm	[Yoshie 01]
pillar microcavity	3× + ab	970 nm	[Moreau 01]
photonic crystal cavity	(9×)	950 nm	[Happ 02]
pillar microcavity	5.8× + ab	855 nm	[Pelton 02]
pillar microcavity	5× + ab	920 nm	[Vučković 03]
pillar microcavity	$str.\ cpl.$	937 nm	[Reithmaier 04]
photonic crystal cavity	$str.\ cpl.$	1 200 nm	[Yoshie 04]

2

Experimental techniques

2.1 Photonic crystal fabrication

The fabrication of good quality PhC devices is a challenging process. Each step needs a careful optimization and constant monitoring. The goal is to obtain an almost perfect crystal of air cylinders with smooth vertical sidewalls of controlled diameter. Optical microscopy and SEM, are useful in checking the outcome of each fabrication step. However, the ultimate assessment for quality comes from cavity related properties, as cavity mode wavelength and spectral width.

I worked in collaboration with A. Gerardino and M. Francardi from the Institute for Photonics and Nanotechnologies (IFN-CNR) in Roma for the fabrication process. They did most of the clean room processing, based on my cavity designs, except for the etching of the PhC holes and the release of the membrane, which I performed at EPFL.

Membranes

The samples were grown at EPFL by **M**olecular **B**eam **E**pitaxy (MBE) (see section 1.1.3). The vertical structure of the sample can change, depending on the project, but they usually contain a GaAs layer with embedded QDs on top of a wide $Al_{0.7}Ga_{0.3}As$ sacrificial layer. The sacrificial layer is removed at the end of the process to produce a free standing GaAs membrane. The usual membrane thickness is 335 nm for room temperature and 320 nm for low temperature samples. It can support two guided modes at 1300 nm, but only the fundamental mode has a significant overlap with the QDs and is affected by the PhC band gap. The $Al_{0.7}Ga_{0.3}As$ sacrificial layer is usually 1.5 µm thick and, once removed, the empty space is wide enough to prevent the leaking of the evanescent part of the guided mode into the GaAs substrate. The processing steps are listed below:

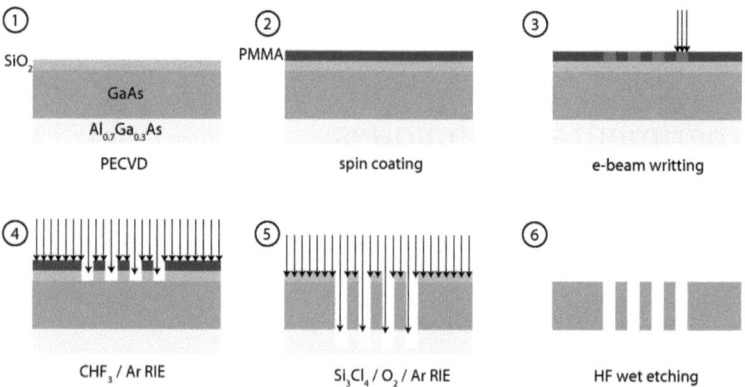

Figure 2.1: Processing steps of a photonic crystal on a GaAs membrane.

1. Deposition of a 150 ∼ 200 nm thin layer of SiO_x by **P**lasma-**E**nhanced **C**hemical **V**apor **D**eposition (PECVD) on top of the sample. I performed the first depositions at EPFL, but then the group in Roma decided to do them to have more control and flexibility in the rest of the mask writing process.

2. Spin coating of a 200 nm thick layer of **P**oly(methyl methacrylate) (PMMA) — also known as plexiglass. It serves as a positive resist for electron beam lithography, since the exposed area is more soluble to developers.

3. Writing of the mask onto the PMMA with electron beam lithography (Leica EBPG5-HR). The minimum exposed feature is of approximately 20nm. The PMMA is then developed in a solution of methyl isobutyl ketone and isopropyl alcohol.

4. Mask transfer to the SiO_x by CHF_3 **R**eactive **I**on **E**tching (RIE). The sample is then sent to Lausanne for further processing.

2.1. Photonic crystal fabrication

5. Drilling of the PhC holes into the GaAs layer. We need highly directional etching to transfer the mask pattern vertically all the way through the membrane. Since wet etching is not well suited to this task, we used RIE (see figure 2.2). The reactor consists of a vacuum chamber with two plate electrodes between which a RF oscillating electromagnetic field can be applied. The top electrode and the chamber are grounded whereas the bottom electrode is electrically isolated. Gas can enter the chamber at a controlled flow rate, and the pressure is adjusted by a valve. The samples are placed on a graphite plate that can be transfered from the load-lock onto the bottom electrode in the etch chamber. (see figure 2.2)

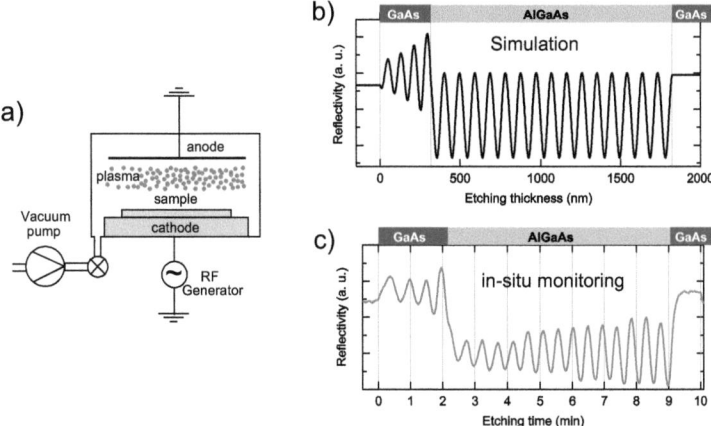

Figure 2.2: (a) Sketch of the RIE chamber. (b) Picture of the etch chamber during the process. (c) Simulated and (d) measured reflectometry performed on a unmasked sample. We clearly recognize the etching of the 320 nm GaAs membrane (0 to 2'15''), of the 1.5 μm $Al_{0.7}Ga_{0.3}As$ layer (2'15'' to 9') and finally of the GaAs substrate.

The strong RF field ionizes the gas molecules and creates a plasma. The electrons are following the oscillating field and some of them are hitting the walls of the chamber, the top electrode or the graphite plate. In the last case, since the plate is DC isolated, they accumulate and build up a strong electric field. As the electrons are repelled by this negatively charged electrode, there are fewer collisions in this region. This is visible immediately above the graphite plate as a darker layer relative to the intense glow of the plasma (figure 2.2b).

On the other hand, the heavier positive gas ions do not respond to the RF field but are accelerated vertically toward the negatively charged electrode where the sample is laying. The anisotropy of the etch has its origin in this vertical delivery of the reactive gas atoms. They are adsorbed on the unmasked surface of the sample and form volatile species whose removal is in turn enhanced by the ions bombardment.

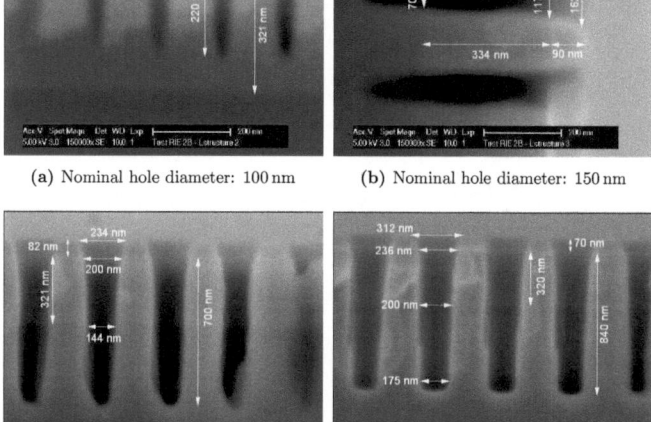

(a) Nominal hole diameter: 100 nm   (b) Nominal hole diameter: 150 nm

(c) Nominal hole diameter: 200 nm   (d) Nominal hole diameter: 250 nm

Figure 2.3: RIE performed on 17/06/2006 on test sample RIE2B. The etch duration was 10 minutes using the parameters listed on table 2.1. Holes with nominal diameter 50 nm did not etch completely through the SiO_2 mask and are not shown here.

Before performing the dry etching of the PhC holes, the sample is usually checked with SEM or optical microscopy. Then, depending on the surface condition, it is cleaned with acetone, isopropanol and/or O_2 plasma (60 W, 1∼2 minutes) to remove organic

2.1. Photonic crystal fabrication

contaminants. The sample is then loaded in the RIE. Once the desired vacuum is reached ($< 3 \cdot 10^{-7}$ mbar) we start the etching with the parameters listed in table 2.1.

$SiCl_4$	O_2(sccm)	Ar	Pressure (mTorr)	Power (W)
25	50	3	10	200

Table 2.1: Etching parameters on the Oxford RIE.

Even though the etch rate depends strongly on the hole diameters, as illustrated in figure 2.3, it is always a good idea to monitor it in situ by using thin film reflectometry on a unmasked area or, as we most often did, on another bare sample (see figure 2.2d).

6. We use diluted HF to remove the remaining SiO_2 mask and to obtain a free standing membrane by etching away the $Al_{0.7}Ga_{0.3}As$ sacrificial layer below the PhC. We chose different concentrations ($< 5\%$) and etching times (< 5 minutes) depending on the goals and on the available chemicals at EPFL or TU/e. We observe an underetch of $3.6\,\mu m$ for 3% HF during 5 minutes (figure 2.4(a)-(b)) and of $1.7\,\mu m$ for 3% HF during 2 minutes (figure 2.4(c)-(d)). Since HF is a hazardous substance, we tried to work with small volumes in the range of 10 to 20 ml per sample. This approach has an impact on the etching rate of large samples due to the depletion of active etchant species. To obtain consistent under-etched profiles, we used a fresh solution for each separate sample. We noticed that the purity of the water, as well as the temperature of the solution have consequences on the etching speed and quality.

Optical microscopy or SEM (≥ 10 kV) can be used to measure the extent of the under-etched region, as seen on figure 2.4.

7. The drying of single membrane samples does not represent a real challenge as long as the distance between surfaces is large enough (see figure 3.41). The main idea is to replace the water with another solvent of lower viscosity to avoid the collapse of the membrane by sticking. We usually use hot isopropanol, around its boiling point ($82.3\,°C$). We either hold the sample in the vapor until all the water has dropped out, or directly dip the sample in the solvent and hold it in the vapor. We then keep it above the hot plate to evaporate the remaining isopropanol droplets condensed on the surface. Since the last drop usually contains all the impurities that are left behind

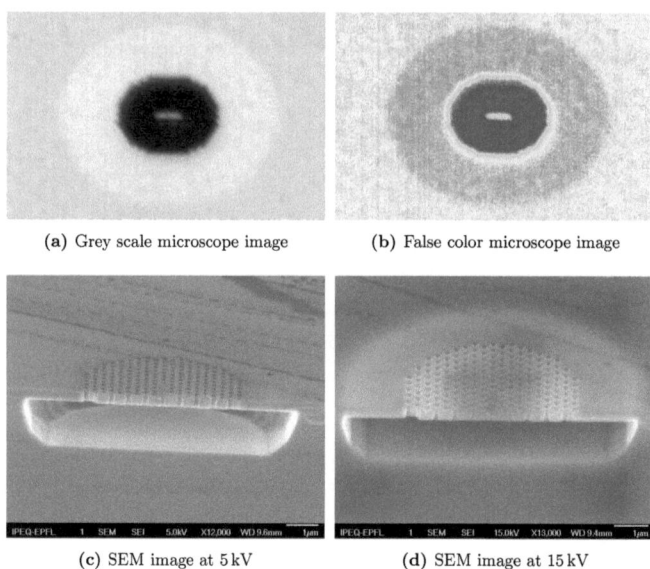

Figure 2.4: The extension of the under-etching can be measured by optical microscopy (a)-(b) or by SEM (c)-(d). In the first case, the contrast depends on the reflection of the light at the interfaces, whereas in the second case, it depends on the penetration depth of the electron beam into the material and thus on its acceleration voltage. For voltages higher than 10 kV, the underetched area is seen as a brighter halo around the photonic crystal region (d). Cleaving of the sample is thus not necessary.

after complete evaporation, we are cautious not to let it dry on a patterned region of the sample.

Paths for improvement

Mask writing Recently, cavities with quality factors $Q \approx 7 \cdot 10^5$ have been demonstrated in GaAs membranes [Combrié 08]. The authors compare this result to their previous record of $Q \approx 2.5 \cdot 10^5$ and stress that the only cause of improvement is a new e-beam writer

2.1. Photonic crystal fabrication

with higher resolution. This accentuates the importance of the e-beam writing step. On our samples, we sometimes observe distorted hole geometries and irregular positioning of the holes due to beam instabilities, stitching problems when the sample is physically moved to another writing block, merging of neighboring holes due to proximity effect, as well as photonic crystal anisotropies whose symptom is clearly visible in photoluminescence spectra as a splitting of degenerated modes. This anisotropy comes either from different scanning steps in x and y directions, or more probably from the deflection of the beam at the border of a writing block. In this case, the angle under which the beam is reaching the sample will cause an astigmatism of the writing spot, as illustrated schematically on figure 2.5.

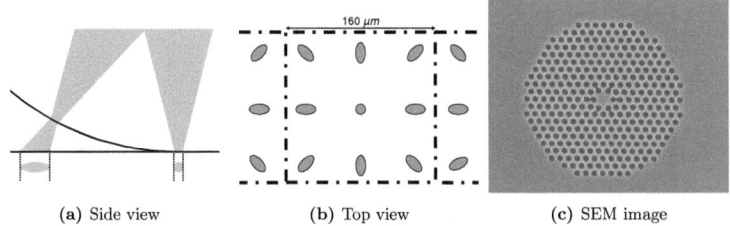

(a) Side view (b) Top view (c) SEM image

Figure 2.5: (a)-(b) The electron beam is focused to form a round spot of 5 nm at the center of a writing block. When deflected from the center, it becomes defocused and astigmatic. (c) On this SEM image, the holes are elliptic with an elongation of 2%. The merging of holes visible at the center of the photonic crystal is due to proximity effect.

A practical path for improvement would be to use smaller structures, place them at the center of the writing block, and perform the mask transfer when there is less electromagnetic noise, typically at night. We already observed that the latter gives better results. Proximity effects can also be avoided by adjusting the dose of exposition when structures are getting closer, as for modified L3 or H1 cavities (figure 2.5c)

Hole verticality The verticality of the holes sidewall can also be improved. When we observe the holes on figure 2.3, we notice that the top aperture is funnel shaped and that the rest of the hole is straight. This comes from the shape of the SiO_x mask. A way to circumvent it would be the growth of a ~ 50 nm $Al_{0.7}Ga_{0.3}As$ layer on top of the sample. During the dry etching, the incurved part would remain in this layer, that will be wet etched away with the release of the membrane.

Another approach to improve the holes sidewall would be to perform the etch with Inductively Coupled Plasma RIE. The operating principle of ICP is very similar to RIE: a RF magnetic field is responsible for high density plasma generation, whereas a separate RF bias is applied to the base plate to create directional electric fields near the substrate. This provides control on the current of ions flowing toward the sample and thus of the anisotropy of the etch profile. I performed the first tests at EPFL and Matthias Skacel carries on this project at TU/e (figure 2.6).

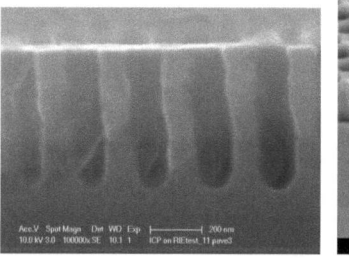 

Figure 2.6: SEM view of ICP dry etches performed at EPFL with BCl_3/N_2 plasma (left) and at TU/e (right) using Cl_2, H_2 and Ar.

2.2 Measurement setups

The optical characterization of single QD emission requires a photoluminescence setup with a good spatial and spectral resolution to detect single excitonic lines.

2.2.1 Micro-photoluminescence setup

Figure 2.7 presents the micro-photoluminescence setup (micro-PL) we used for some of our experiments. It was built by C. Zinoni in the framework of his thesis work at EPFL. The micro-PL setup is equipped with a continuous helium flow cryostat. The temperature of the sample was measured to be higher than that of the heat exchanger, and does not go below 10 K. We used a fiber-coupled pulsed laser, with pulse width of ∼50 ps and emission at 750 nm for excitation in the GaAs. The micro-PL signal is collected through the same

2.2. Measurement setups

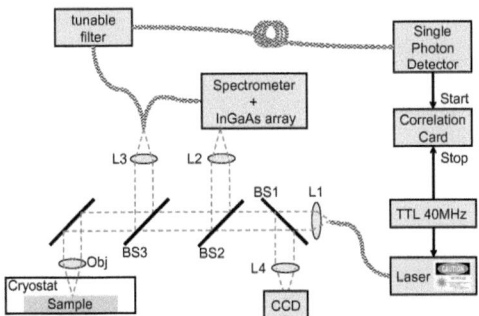

Figure 2.7: Sketch of the micro-PL setup. Obj is the microscope objective, Lx are the lenses, and BSx the beam splitters or dichroic mirrors, depending on the wavelength of the laser. The yellow links represent single mode optical fibers.

microscope objective (NIR, 100×, NA=0.5) and then coupled to a single mode fiber (SMF 28) or sent directly in free space to a spectrometer (Horiba-Jobin-Yvon FHR 1000) equipped with a LN$_2$ cooled InGaAs array with a maximum resolution of $\sim$ 30 μeV ($\sim$ 0.04) nm, depending on dispersion grating selected. The single mode fiber acts as a pinhole and reduces the collection area on the sample. PL spectra recorded through the fiber entrance of the spectrometer have less background luminescence from QD not coupled to the cavity.

The fiber can be coupled to a tunable filter and to a single photon detector to perform time-resolved spectroscopy. At the beginning of this thesis, we used a InGaAs avalanche photodiode (APD), but then switched to a superconducting single-photon detector (SSPD).

2.2.2 Superconducting Single-Photon Detector

A SSPD is basically a narrow thin wire of superconducting material (NbN) biased close to its critical current. When a photon is absorbed, it creates a local hot spot in the wire. The current continues to flow through the wire, avoiding the hot spot. But since the wire is narrow, the critical current density for superconductivity is exceeded, and the full section of the wire becomes resistive. This is detected as a voltage pulse in the transmission line. The detection quantum efficiency and the dark noise counts depend on the applied bias. It is possible to reach 10% efficiency for 1 $\sim$ 2 Hz of noise, which is a huge improvement as compared to InGaAs APDs. More details on SSPDs can be found in [Gol'tsman 01], and specifically for our detectors in [Korneev 07].

2.2.3 Cryogenic probestation

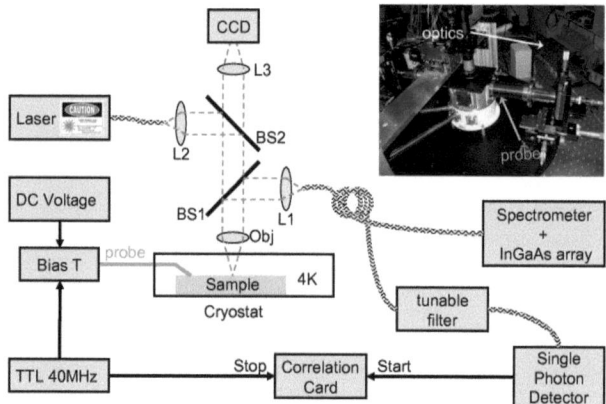

Figure 2.8: Sketch of the cryogenic probe station. The labels are the same as for figure 2.7. The orange link from the bias-T to the sample represents the microwave probe.

Electro-optical characterizations (see sections 3.9 and 5.2) were performed inside a cryogenic probe station (Janis). The electrical contact is realized by a cooled 50 Ω microwave probe attached to a micro-manipulator, and connected by a coaxial line to the room temperature circuitry. Continuous bias can be supplied by a low noise voltage source through the DC port of a bias-T. The AC port of the bias-T can be used to drive the device with pulsed excitation, for example to record time resolved PL. The electroluminescence is collected through a long working distance objective (MPlan, 10×, NA=0.3 or Cassegrain, NA=0.4) and coupled to a single mode fiber. The fiber can be connected to a single photon detector to perform light–current measurements (L–I), or to a spectrometer for recording PL spectra. Optical excitation can be performed with a fiber coupled laser, focussed on the sample through the same objective.

The collection efficiency is not as good as for a conventional micro-PL setup, which makes experiments more challenging.

2.2.4 Tri-axial micro-photoluminescence setup

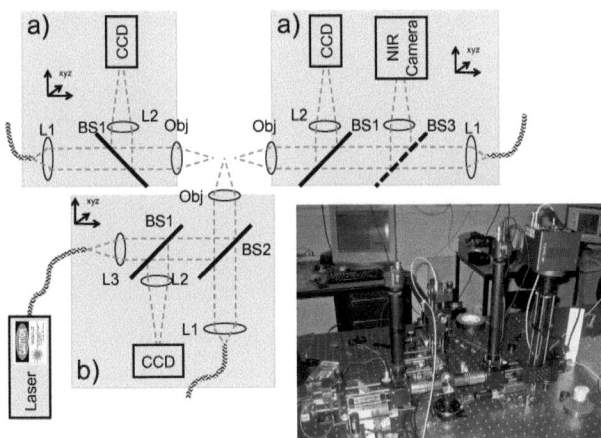

Figure 2.9: (a–b) Sketch of the constitutive elements of the tri-axial micro-PL setup. It is made of three independent units, 2×(a) and 1×(b), mounted on xyz piezoelectric stages. The two units (a) can be used for either the excitation of the PL or the collection of light in transmission experiments, and the unit (b) for micro-PL experiments. (c) Picture of the tri-axial micro-PL setup in room-temperature configuration. The luminescence of QDs in a waveguide is excited with unit (b) and collected from the top of the sample (unit b) and from both cleaved facets (units a) simultaneously.

The tri-axial micro-PL setup (Triax) is the combination of a conventional micro-PL (figure 2.9b) and a transmission measurement setup (2× figure 2.9a). It can be used to excite the PL from the top of a waveguide and collect the light from the two cleaved facets. The picture on figure 2.9 shows the setup in this configuration.

I designed the Triax setup to be versatile and optimized for signal collection. Each unit is mounted on a piezoelectrically driven xyz-stage (Newport 562-Ultralign equipped with NanoPZ actuators) and can be used separately. The light is guided to and from the blocs with single mode fibers: SMF-28 for the horizontal paths, and SM 980, SM 780HP, or SM 600 depending on the laser excitation wavelength, for the vertical path. In micro-

PL configuration, the excitation and collection paths are superposed through a pellicle beamsplitter (BP133) with R:T ratio of 50:50 at 600 nm, 40:60 at 750 nm, and 20:80 at 1300 nm. A large portion of the excitation is lost, but this choice benefits to the collection efficiency.

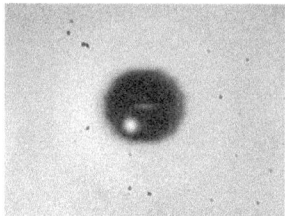 

Figure 2.10: (left) Monochrome image of a L3 cavity with lattice parameter $a = 400$ nm, showing the under-etched region and the excitation laser spot. (right) Close view of the Triax setup in the low temperature configuration.

The light is focussed on the sample with a microscope objective (two M Plan Apo NIR 100×, NA=0.5, WD=12.0 mm; one M Plan Apo NIR 50×, NA=0.42, WD=17.0 mm). At 1300 nm, the objectives have $\sim 60\,\%$ transmission.

We use uncoated pellicle beamsplitters (8:92) to direct the light to low-light CCD cameras (Watec 902H2). A useful feature is its ability to see intense 1300 laser light focussed on the sample, probably through two-photon absorption, which helps in the alignment procedure. It is also possible to use a InGaAs camera (blue box on figure 2.9) to record infrared light. The illumination is provided through blue-green or IR LEDs. The $1/e^2$ diameter of the laser spot, for a wavelength of 750 nm and the 100× microscope objective is of approximately 1.5 µm. (see figure 2.10)

The setup can also be operated in a low temperature configuration in conventional micro-PL configuration, but also in orthogonal excitation-transmission geometry by using a cube window. With the cube window, the temperature of the sample is around 40 K, since there is no radiation shield. I am currently working on improving this feature.

3

Tuning

Equation (1.10) indicates three paths to obtain a maximized rate enhancement in case of emitter to cavity coupling:

The Purcell factor F_p can be maximized by carefully selecting cavity designs with low modal volume $\mathcal{V}_{\text{mode}}$ and high quality factors Q. For example, point defect cavities, such as H1 or L3, can be optimized to reduce losses. Other cavity designs have been developed along single line-defects, i.e. photonic crystal waveguides, either by modifying the lattice parameter of the photonic crystal as in double-heterostructure cavities or by locally modifying its width.

Table 3.1: Calculated and measured Q factors for different photonic crystal cavities with modal volume $\mathcal{V}_{\text{mode}} \approx (\lambda/n)^3$.

Cavity type	$Q_{\text{simulation}}$		Q_{measured}	
modified L3	$3 \cdot 10^5$	[Akahane 03]	$1.1 \cdot 10^5$	[Galli 09]
modified H1	$1.4 \cdot 10^6$	[Tanabe 07]	$4 \cdot 10^5$	[Tanabe 07]
disordered line defect	—		$1.5 \cdot 10^5$	[Topolancik 07]
double-heterostructure	$2 \cdot 10^7$	[Song 05]	$2 \cdot 10^6$	[Song 07, Noda 07]
local width modulation	$\sim 10^8$	[Kuramochi 06]	$1.34 \cdot 10^6$	[Notomi 07]
modified heterostructure	$\sim 10^9$	[Tanaka 08]	—	

Some of the designs, along with the expected and measured Q factors are summarized on table 3.1. The modal volume V_{mode} is on the order of $(\lambda/n)^3$. These impressive results were usually measured by transmission measurements on empty Si cavities. On GaAs membranes, Hermmann et al. measured $Q = 1.4 \cdot 10^5$ on an empty double-heterostructure cavity [Herrmann 06] and Combrié et al. reached $Q = 7 \cdot 10^5$ using the local width modulation design [Combrié 08].

In our case, we do not expect to observe such high Q factors, since we use the luminescence of QDs as an internal light source. A photon emitted in the cavity mode has a probability to be reabsorbed by other QDs, and is thus "lost" by the cavity, thereby decreasing the measured Q. As pointed out by Hennesy et al. , the number of QDs in a cavity has a negative influence on Q, and they reached $Q = 12\,000$ to $40\,000$ for modified L3 cavities containing exactly one QD [Hennessy 07]. In our case, the samples with low density QDs have on average several ($5 \sim 10$) QDs in the cavity. We reached our record value of $Q = 16\,500$ on a modified L3 design.

The spatial overlap can be improved by knowing the position of the QDs, for example by measuring their position and arranging the cavity around them [Badolato 05]. Another approach is to use self-assembled QDs grown on pre-patterned substrates. However, they usually show poor optical properties. Integration with other types of deterministically grown QDs was also demonstrated [Gallo 08]. In this work, we adopted a statistical approach: we built a large number of cavities and patiently measured them to estimate the coupling.

The frequency matching can be optimized by acting on the mode or on the QD emission. Even though the size dispersion of the QDs is relatively small (see figure 1.1), it is not possible to control exactly their emission wavelength. Similarly, even when carefully choosing the right parameters for the photonic crystal, the frequency of the mode can be perturbed by imperfections in the fabrication process. To account for them, we usually repeat the same structure with different lattice parameters. Various post-processing techniques have also been proposed. For example wet chemical digital etching acts on the Q and on the mode energy by increasing the filling factor F [Hennessy 05], atomic force microscope (AFM) nano-oxidation can be used to overlap degenerated modes [Hennessy 06], or nano-infiltration in selected holes of the photonic crystal [Intonti 06].

3.1. Lithographic tuning

The ability to actively tune the mode or the QD frequency gives control on the degree of coupling, and thus on the lifetime of the emitter through the Purcell effect. The properties of the dielectric environment or of the emitter can also be dynamically tuned, for example with the temperature [Yoshie 04, Englund 07], by creating free carriers with an ultrafast laser [Fushman 07], by deposing material on the cavity surface [Mosor 05], by infiltrating liquid crystals in the holes [Leonard 00], by approaching a dielectric object close to the cavity [Märki 06, Hopman 06, Mujumdar 07], etc.

The ideal tuning strategy should modify only the wavelength of the mode or of the QD, without lowering the Q or broadening the exciton linewidth, as this would have an uncontrollable effect on the measured lifetimes, (see equation 1.14)

In this chapter, we will investigate different tunning strategies. In section 3.1 we will report on our results in the optimization of photonic crystal cavities, in sections 3.2, 3.4, and 3.7 we will review the effects of temperature on the mode and on the QDs energy, in section 3.3, 3.5, 3.6, and 3.8 we show the effects of different approaches to modify the dielectric environment of the cavity mode, and finally in section 3.9, we demonstrate electrical control of the energy of the QDs.

3.1 Lithographic tuning

Photonic crystal is a versatile tool to probe the light-matter interactions. Every parameter can be adjusted to ones needs: the dimensionality, the type of crystal structure, the shape and size of the "atoms", the introduction of defects or disorder in the structure, etc.

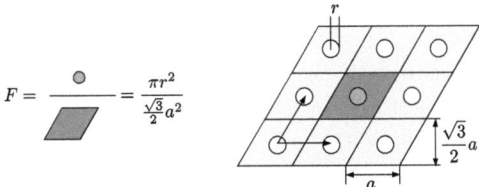

Figure 3.1: Definition of the filling factor. $F = 35\%$ corresponds to $r \simeq 0.31 \cdot a$

Triangular lattice of holes

Since we decided to work with photonic crystals in membranes, we could therefore rule out lattices composed of dielectric columns in air. Then, as the light emitted by our QDs is TE-polarized, we selected a triangular lattice of holes drilled in the membrane, since it offers a relatively wide band gap for this polarization as compared to square ones. We also used round atoms, as the small dimensions of the structures would make other shapes challenging to fabricate reproducibly. A dispersion diagram of such a triangular lattice is provided in figure 3.2a. It was calculated for TE polarization, for empty holes ($n_{holes} = 1$) in a GaAs membrane ($n_{eff} = 3.0912$) and a hole filling factor of 35%. We can observe a relatively wide full photonic bandgap, in which guided light can be trapped in all directions.

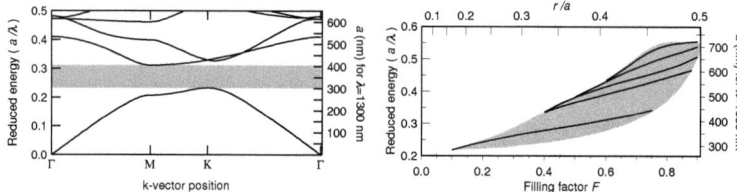

Figure 3.2: (left) Dispersion diagram of a triangular photonic crystal of empty holes in a semi-conductor ($F = 35\%$) with effective refractive index $n_{eff} = 3.0912$, corresponding to 1300 nm light propagating in a 320 nm thick GaAs membrane at room temperature. The right axis gives the corresponding lattice parameter value. (right) TE full photonic bandgap (shaded area) and H1 cavity modes energy (lines) as a function of the filling factor F in this photonic crystal.

Filling factor F

The next parameter to be set is the filling factor. It is defined in figure 3.1 as the ratio of air hole to unit cell area. The filling factor has a direct influence on the size and position of the bandgap, as illustrated on figure 3.2. It also affects the energy of cavity modes. For example an increase in F results in a blue shift of a H1 cavity mode. An intuitive explanation is that, as the holes grow larger to the detriment of the dielectric material, the effective volume of the cavity mode is reduced, thus resulting in the observed blue shift.

Figure 3.3 illustrates the effect on a L3 cavity. In this case, the filling factor change was not designed directly: we wanted to determine the number of hole shells needed to obtain

3.1. Lithographic tuning

a good confinement. During the electron-beam lithography, the structures with more holes, for which the beam needs to stay for a longer time, were slightly over-exposed, resulting in larger holes. This is known as the proximity effect. We clearly see a blue shift of all the peaks, and also notice, by looking at the distance between them, that they move at different speeds. This can be explained by the different sensitivity of the mode profile to the hole diameter due to their different geometry (see figure B.4).

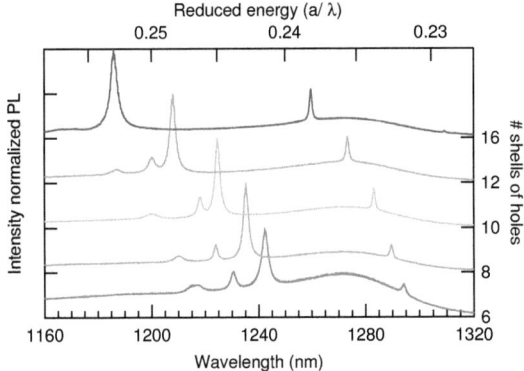

Figure 3.3: Illustration of the mode shift induced by a variation of the filling factor on a L3 cavity. F increases along with the number of shell holes. The modification in the hole diameter of these nominally identical cavities is due to the proximity effect in the e-beam writing of the mask, owing to the increasing number of hole shells around the cavity (right axis).

Intuitively, modes close to the band edges are less confined by the photonic crystal. The decrease in reflectivity causes a degradation of the quality factor Q. As the width of the band-gap increases linearly[a] with the filling factor up to $F = 70\%$, it seems reasonable to choose it as large as possible. An obvious limitation comes from the technological point of view. On one hand, large holes will make the membrane more fragile. They will also increase the surface to volume ratio, thus statistically creating a larger number of surface defects, i.e. increase non-radiative recombinations. On top of that, as the profile of the holes in the SiO_x mask is not vertical, larger holes effectively reduce the mask thickness, thus reducing the quality of the hole profile in the photonic crystal. We designed cavities

---
[a]The width of the band-gap can be approximated by $(0.328 \pm 0.002) \cdot F - (0.0370 \pm 0.0007)$ for $F < 70\%$.

with different filling factor values and, based on the experimental results, decided to target $F = 35\%$, corresponding to $r/a \simeq 0.31$.

Lattice parameter a

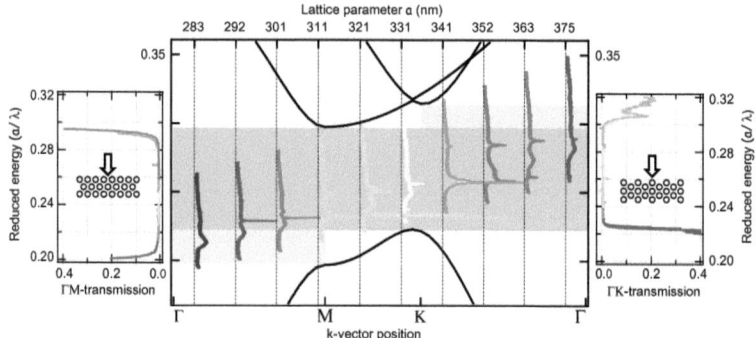

Figure 3.4: Close-up view at the dispersion curves near the band-gap. The center part displays the micro-PL spectra of L3 cavities with different lattice parameters a (top axis) plotted vertically as a function of their reduced energy (a/λ). The left (right) graph shows the measured transmission of light through the photonic crystal along the ΓM (ΓK) direction, illustrating the photonic gap effect.

We usually use QDs with ground state emission around 1300 nm to probe the photonic band-gap in a so-called internal light source configuration. It has a gaussian shape with a full width of $\sim$50 nm at half maximum, and is thus not broad enough to cover the whole band-gap. We could imagine embedding QDs with different emission wavelength to broaden the spectrum. However, as photonic crystals are in principle totally scalable, a given reduced energy (a/λ) can also be reached by sweeping the lattice parameter a. This is illustrated in figure 3.4 for ten L3 cavities with lattice constants a in steps of $\sim$10 nm. The spectra are rotated 90 degrees from the usual way of presenting them, and are plotted versus the reduced energy. We can clearly follow the QD ground state emission across the gap, and we notice three modes appearing at given energies, increasing in intensity as they match the QDs energy. The same technique was applied to measure the transmission of light through

3.1. Lithographic tuning

the photonic crystal along the ΓM and ΓK direction. The luminescence of the QD was excited from the top of the sample, but in this case, it was collected from a cleaved facet, some ten μm away of the photonic crystal (see for example [Ferrini 02]).

If, on the other hand, we are not interested in probing the whole photonic gap, but we want to couple single QDs to a given mode, we will use a smaller sweeping range approximately ±15 nm around the mode energy to account for imperfection in the fabrication process and to provide several starting points for the post-processing tuning techniques presented in the next sections of this chapter.

Choice of a cavity design

The choice for all the parameters fixed so far was mostly dictated by the nature of the light source (epitaxial growth, wavelength) and of the III/V technology. We have now to decide which structure we want to draw in the photonic crystal, keeping in mind our goal of light-matter interaction. As this requires resonators with high finesse and low modal volume, we investigated different designs. We concentrated our efforts on point defect cavities and explored different layouts with few missing holes as reported in figure 3.5 Some cavity designs are similar to those previously reported in the literature (e.g. H1, H2, L3), others are original. Not all of these *exotic* cavities showed good Q factor, nevertheless they proved to be useful in validating the SNOM setup[b] and the 3D FEM Maxwell solver[c].

For example, the O2 cavity can be used for infiltration of liquids (active/passive) in the center of the cavity and the D2 has two modes relatively close in energy with orthogonal polarization that can be used for cavity coupling applications as demonstrated by [Vignolini 09].

Fine tuning of the cavity design

The properties of the cavities can be fine tuned by slightly altering their shape, for example by adjusting the size and position of some of the surrounding holes. This can be used to improve the far-field pattern, as presented in section 4.1, or to increase the quality factor Q by gently confining the mode, as explained in [Akahane 03, Akahane 05].

We designed a sample with modified L3 cavities. The first holes on each side are slightly pushed outwards, and their size is also reduced. The results are displayed on figure 3.6 where

---
[b] Group of Prof. Massimo Gurioli, European laboratory for non-linear spectroscopy, University of Florence
[c] Group of Prof. Bernd Witzigman, Integrated Systems Laboratory, ETH Zürich

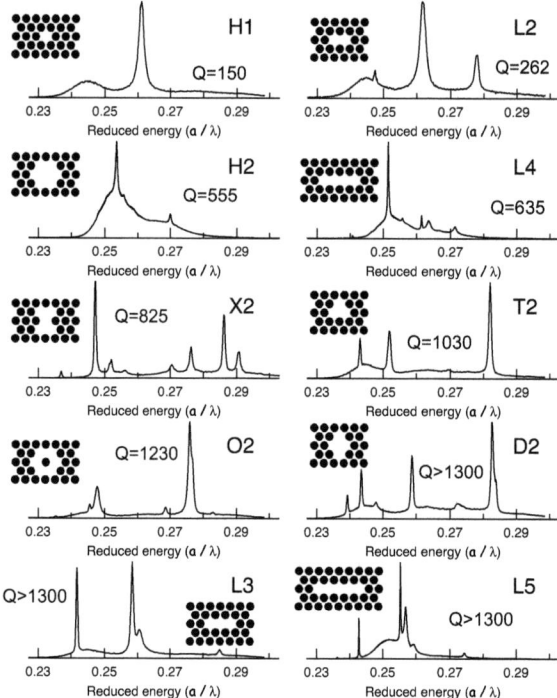

Figure 3.5: Various point defect cavity layouts realized on our early samples, with corresponding micro-PL spectra and measured quality factor Q on the thinnest mode.

we observe higher Q factors for displacements between $0.125 \cdot a$ and $0.150 \cdot a$ together with first hole radius between $0.4 \cdot r_0$ and $0.5 \cdot r_0$. We then produced several batches of samples with parameters in the optimal range. The quality factor varies from sample to sample and even between cavities with the same nominal parameters on the same sample. However, samples with a high density of QDs have smaller Q values than low density samples: reabsorption by QDs degrades the lifetime of the cavity and hence its Q. Our highest Q factors for modified L3 cavities were reaching 16 500 on a sample with low density quantum dots ($\sim 10\,\mu m^{-2}$).

3.1. Lithographic tuning

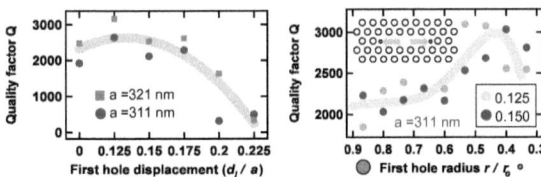

Figure 3.6: Effect on the Q factor of (a) moving the first lateral holes outwards while keeping their radius constant. The blue dots (red squares) are for a lattice parameter $a = 311$ nm ($a = 321$ nm). (b) For a displacement of $d/a = 0.125$ (green dots) and $d/a = 0.15$ (blue dots) we reduced the diameter of the first lateral holes (see inset) in the range $1 \geq r/r_0 \geq 0.3$. The thick grey line is a guide for the eye.

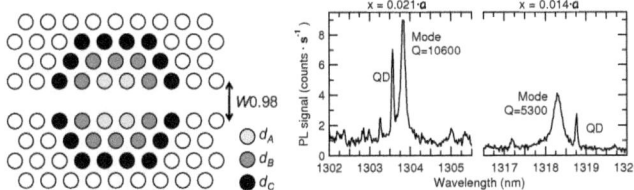

Figure 3.7: (a) Design of a cavity based on the local modulation of the width of a $W0.98$ line defect. The holes forming the modulation are moved perpendicularly away from the line defect by $d_A = x$, $d_B = 2x/3$, and $d_C = x/3$, where x is the tuning parameter. (b) PL spectra of two cavity modes with $Q \simeq 10650$ for $x = 0.021 \cdot a$ (left) and $Q \simeq 5600$ for $x = 0.021 \cdot a$ (right), measured at liquid helium temperature on a sample with low density QDs.

Another type of nanocavity, which is not based on a single defect but on the local width modulation of a line defect has been proposed in [Kuramochi 06]. Basically, this design can be seen as a step forward in the gentle confinement of a mode, since it simply removes the holes in ΓK and lets the field expand naturally in this direction. The lateral size of the waveguide is reduced to $0.98 \times 2a$ (i.e. a W0.98 waveguide) and the confinement is provided by slightly shifting three layers of holes perpendicularly away from a single line defect, as shown in figure 3.7a. The first layer is shifted by $d_A = x$, the second by $d_B = 2x/3$, and $d_C = x/3$ for the third one, where x is the tuning parameter. Simulations indicate a maximal value of Q for $x = 0.021a$. The second part of this figure displays PL spectrum of the mode we obtained for two cavities with $x = 0.021a$ (right) and $x = 0.014a$ (left) with low density QDs at liquid helium temperature. Spectra were also recorded at higher temperature and with stronger excitation to discriminate the cavity resonance from QD excitonic transitions. This is a very promising results, as we measured Q factors in excess of 10^4 for the first batch of cavities processed. This also illustrates the sensitivity of the cavity mode to the tuning parameters, and consequently to fabrication disorders.

3.2 Temperature tuning

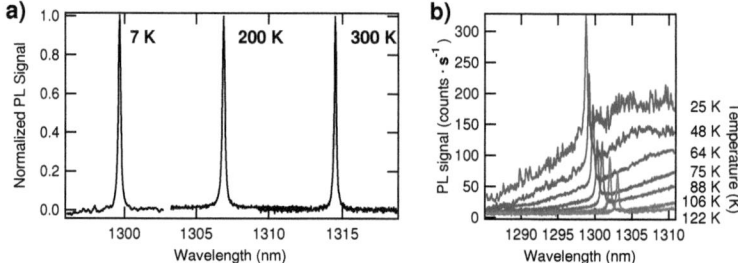

Figure 3.8: (a) Set of three spectra recorded at 300 K, 200 K, and 7 K showing a blue shift of the cavity mode of 15 nm while cooling from room to liquid helium temperature (b) The range of temperature is between 122 K and 25 K. We observe a slight quality factor decrease (∼10%) explained by the reabsorption in the blue shifted QDs.

The study on single QD to cavity coupling is usually performed at cryogenic temperatures to avoid the effect of phonon broadening of the QD line-width. We can use the temperature dependence of the emission energy of the QDs states to tune them in resonance with a cavity mode. For example, a sample with ground state emission of 1400 nm at room temperature will blue shift to 1300 nm at 4 K[d]. However, as we want to keep narrow

---

[d] We usually use a liquid helium continuous flow cryostat and glue the sample with thermal grease (Apiezon® N) on the cold finger. As the temperature of the cryostat is measured at the heat exchanger, it can differ significantly from the temperature of the sample. In addition, the thermal conductivity of the bonding depends on the quality of attachment (amount of grease, number of thermal cycles), and we

3.2. Temperature tuning

excitonic lines and thus work below ~50 K, we have a tuning range of approximately 2 nm (see figure 3.9d).

The refractive index of GaAs is also temperature dependent. This has an influence on the photonic crystal cavity modes, as displayed on Figure 3.8a where we observe a 15 nm blue shift of the mode while cooling from 300 K to 7 K. Thus, both QDs and mode red shift their energy with temperature. But, as the mode shift is one order of magnitude smaller than for QDs, we can still use temperature to tune them in resonance.

Figure 3.8b illustrates the shift of both the cavity mode and the ground state emission of the QDs from 122 K (bottom curve) to 25 K (upper curve). The intensification in photoluminescence is due partly to the blue shift of the QD emission, but also to the increase of the radiative efficiency due to the reduced thermal escape. We measure a gradual decrease in the quality factor Q of the mode from 4000 to 3500 as the QD emission increases its overlap with the mode. This can be explained by resonant absorption in the ground state of the QDs.

We can have a closer look at the interaction between the mode and the QDs as a function of temperature between 19 K and 42 K in figures 3.9a and 3.9b. They both represent the same data set: the curves are shifted vertically for clarity and the axis ranges are different to emphasis the mode (a) or the QD excitonic lines (b). Their respective wavelength shift is reported in figure 3.9d, where we see that the mode shift (thick line) of $\Delta\lambda_{mode} = 0.33$ nm is five times smaller than the $\Delta\lambda_{QDs} = 1.78$ nm of the QDs in this temperature range. This allows us to observe a crossing with an excitonic line at 1307.8 nm for a temperature of 36.2 K. As it is not possible to follow the position of the QD during the crossing, its wavelength was estimated based on the shift of the other neighboring lines. This is represented as the dotted portion of the line on the graph.

It is interesting to look at the spectrally integrated intensity of the mode as a function of the temperature. We observe a maximum intensity around 36 K, the temperature at which the QD line is in resonance with the mode. In this case, the emission is approximately 37% stronger than out of resonance. To accurately quantify this enhancement, we plotted the spectrally integrated intensity as a function of the central position of the mode in figure 3.9c. The shape can be fitted with a Lorentz function, as expected in the case of weak coupling regime. It would be tempting to attribute this light enhancement to the Purcell effect.

However, since we performed this experiment under pulsed excitation with a repetition

---

measured that the sample base temperature is around 10 K to 15 K.

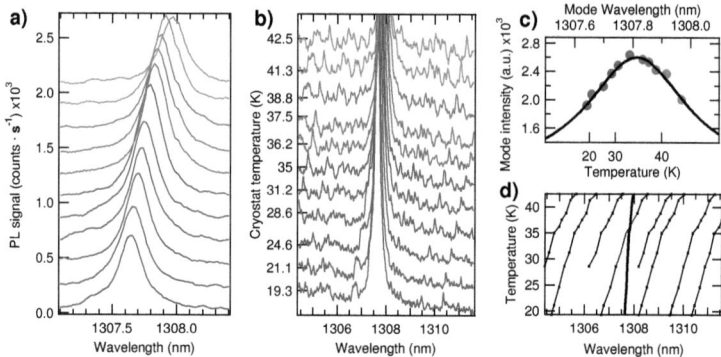

Figure 3.9: Set of spectra recorded at different temperatures ranging from 19 K to 43 K. (a) Close view at the cavity mode. The curves are shifted vertically for clarity. (b) Same as (a) with a broader wavelength range and narrower amplitude scale to reveal the energy shift of single QDs lines. (c) Spectrally integrated intensity of the mode as a function of the temperature. The top axis indicates the central wavelength of the mode. The thick curve is a Lorentzian fit. (d) Wavelength of different QD lines as a function of the temperature. The thick line is the mode and the dotted line is the approximate position of a QD line crossing the mode.

rate more than one order of magnitude smaller than the QD decay rate, this PL spectra do not allow to observe the rate enhancement from the Purcell effect. The intensity increase is rather attributed solely to a better extraction of the photons emitted by the QD: as a result of the coupling, they are funneled out of the GaAs through the cavity mode and thus collected more efficiently. To obtain an estimation of the Purcell effect, we should either measure the dynamics of the QD line, or perform the experiment again under continuous excitation and compare the spectra, as suggested in chapter 2.

Since we need to work at cryogenic temperatures to observe sharp excitonic lines in QDs, temperature tuning is a straightforward method to control the energetic alignment with the mode. However, the useful tuning range is somewhat limited, as increasing temperature also affects the shape and lifetime of the QDs transitions due to phonon broadening along with the thermal activation and non-radiative recombination of carriers.

3.3 Gas deposition at cryogenic temperatures

During our first measurement campaigns, we relied on temperature tuning to move QD and cavity mode on resonance. However, we noticed that even at constant temperature, the wavelength of a cavity mode would change if we measured it at different times during the day. As we tried to understand this phenomenon, we decided to record the position of a mode over a whole day at constant temperature, as reported in figure 3.10, and we observed a constant red shift. After a thermal cycle to room temperature, and subsequent cooling, the mode is reset at approximately its initial energy.

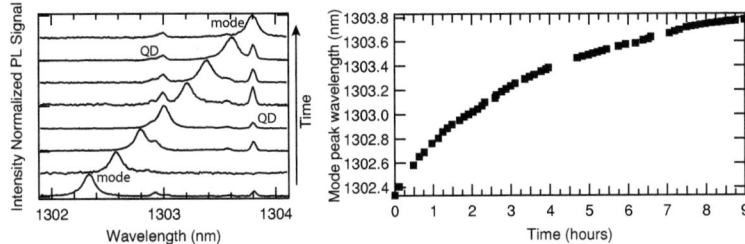

Figure 3.10: Red shift (~ 1.4 nm) of a cavity mode as a function of time for a sample kept at ~ 5 K during 9 hours. (a) Spectra recorded at different times showing the mode drifting to lower energies. The curves are normalized to the mode intensity maximum, which explains the intensity fluctuations of the two QDs at 1303.0 nm and 1303.8 nm. (b) Position of the maximum of the mode as a function of the time.

The idea behind this phenomenon is rather simple and was first described by [Mosor 05]: even though the cryostat is evacuated to $\sim 10^{-6}$ mbar before cooling, there are still gas molecules floating around, either from degassing or from microscopic leaks. As the cold finger reaches temperatures below the freezing point of the gas compounds present in the cryostat, they start being adsorbed on the cold finger. This creates a partial pressure gradient that attracts other molecules to the cold region, as in a cryogenic pump. At the same time, the pressure inside the vacuum system decreases to $\sim 10^{-7}$ mbar, which supports our interpretation.

Since the sample is also at low temperature, gas molecules start depositing on it. They form a thin layer of solid material with an index of refraction higher than vacuum, and than their gaseous counterpart. This decrease in the contrast between dielectric and environment effective index lowers the energy of the photonic air band, and consequently of the cavity resonances: the modes will red-shift.

If we look at the rate with which the mode is shifting, we observe a decrease with time. This can be related to the ongoing adsorption of material inside the holes until they become full but also to the thin layer formed over the surface of the membrane. It increases the effective refractive index of the guided mode, since it interacts with its evanescent field, and thus contribute to the redshift. However, as the field decays exponentially with the distance, the effect has also a limited range of influence.

We also noticed that the initial position of the mode is slightly red-shifted after each thermal cycle. This is attributed to the deposition of a thin grease layer evaporated from the positioning step motors. Rinsing the sample in trichloroethylene, which is very efficient at removing greasy compounds, blue-shifts the cavity resonances to their idle position again.

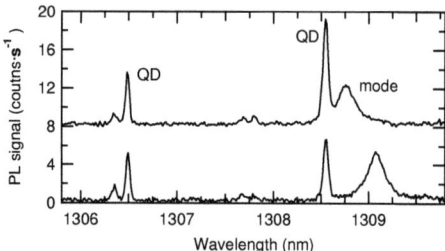

Figure 3.11: Gas deposition provides a clean red shift tuning of the mode, while leaving the QDs lines remain unchanged. We observe a clear increase in the PL of the QD line at 1308.55 nm as it overlaps with the mode.

The time interval between the recording of the two spectra in figure 3.11 is of 3 hours and 15 minutes. During this time, the mode red shifted by only 0.3 nm. The integration time of the second curve was increased to compensate for the reduced collection due to the formation of the condensed gas layer. Still, we can observe a coupling effect, since the integrated luminescence of the QD at 1308.55 nm varies by a factor of 1.8, while that of the other QD lines remains unchanged.

Mode tuning with gas adsorption has the advantage of keeping the QD unperturbed, as compared to temperature tuning. However, the tuning range is rather limited, and the the tuning speed is, so far, the main drawback. We tried to increase the speed by letting in small volumes of air generated between two valves, but we observed a strong degradation in the Q factor of the cavity and in the collected photoluminescence. Following the work done by [Mosor 05], we need to control the pressure of the gas between 0.45 and 0.5 Torr to get a mode shift while keeping a relatively stable Q. Also, the use of Xenon would provide a larger tuning range, up to 5 nm, for the same number of steps, which is equivalent to faster tuning. We are now trying to improve our setup by adding a precision valve and a vacuum chamber and connect it to a pipe inside the cryostat to bring the gas molecules directly near the sample.

3.4 Laser heating and thermal annealing

We worked in collaboration with Armando Rastelli[e] to perform local heating on the photonic crystal by focussed laser beam. The experiments were performed in Stuttgart on samples designed an fabricated in the scope of this thesis.

Description of the setup

The sample is mounted in a Helium continuous flow cryostat and kept at 5 K during the whole experiment. It is then investigated in a conventional micro-photoluminescence setup equipped with a continuous-wave laser with emission wavelength of 532 nm focussed to a $\sim$1.5 µm spot. The laser is used both to probe the photoluminescence of the QDs at low excitation power ($\sim$100 µW) and to heat the sample for the post-processing tuning of the mode or of the QDs (4 to 23 mW).

Local heating for reversible mode tuning

A reference spectrum was taken at the beginning of the experiment (lowest curve in figure 3.12). Then the power was tuned up to the post-processing range for 20 to 30 seconds and lowered to measure the spectrum. It is then increased again to a slightly higher value than in the previous step. For laser powers below 4 mW, there is no observable change in the wavelength of the mode. However, past this value, we start observing mode shifting to the

---
[e]Max Plank Institute für Festkörperforschung, Heisenbergstrasse 1, D-70569 Stuttgart, Germany

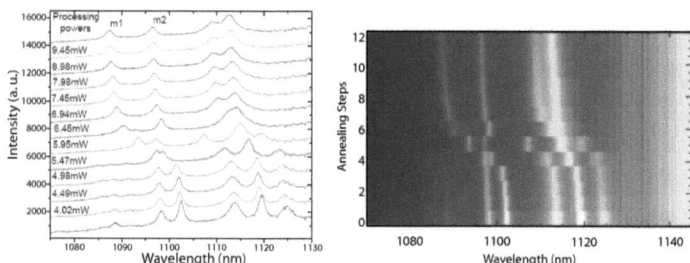

Figure 3.12: Local heating performed on a modified L3 cavity with a high density of QDs. The laser power used to heat the sample is noted at the side of (a) for each step. The modes blue-shift at different speeds, causing a crossing for $m1$ and $m2$.

blue at different rates. This can even lead to mode crossing, as can be seen on figure 3.12 for the modes labeled $m1$ and $m2$.

As we recall from section 3.3, a thin layer of gas deposited on the surface of the sample forms during the cooling. When the temperature reached by the laser induced heating is sufficient, we observe a desorption of those gas molecules and thus a blue shift of the modes. To support this hypothesis, we performed a full thermal cycle of the cryostat until room temperature, and as it reached 7 K again, the modes where back at their initial positions. If we look at figure 3.12, we observe that the mode stop shifting eventually. This is easily explained as a stage where they are no more adsorbed molecules above the cavity.

Estimation of the heating area and of the temperature

The temperature reached by the sample under illumination can be estimated by comparing the wavelength of the mode during the laser heating step (figure 3.13a) and while heating the cryostat (figure 3.13b). The red shift of the mode is caused by the material refractive index n_{GaAs} dependence to temperature. Laser processing powers which start inducing shifts in the modes correspond to temperatures around 200K.

Comparing the behavior of the QD luminescence in figures 3.13a and 3.13b gives an indication of the spatial extent of the laser induced heating. In figure 3.13b, as the cryostat is heated to room temperature, the QD luminescence rapidly shifts across the modes with increasing cryostat temperature. The luminescence intensity decreases, due to thermal

3.4. Laser heating and thermal annealing

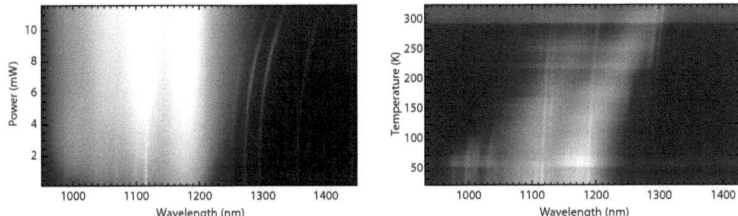

Figure 3.13: (a) Map of the photoluminescence spectra as a function of the laser power recorded during the heating process. (b) Photoluminescence recorded for low laser excitation as a function of the cryostat temperature. The narrow lines correspond to the cavity resonances, and the broad one to the QDs.

escape of the charge carriers out of the QDs. This effect is more pronounced for higher energy states. In figure 3.13a, however, the energy of the QDs remains stable and we observe the sequential filling of ground, first, and second excited states with increasing pumping power. This is a clear indication that the laser induced heating acts locally on the cavity, and that its extent is in any case much smaller than the photoluminescence collection region.

Room temperature thermal annealing

We also performed some experiments at room temperature. Figure 3.14a shows the position of a mode during the annealing process. As expected, the mode clearly shifts to the red under illumination, and back to the blue as the laser intensity is reduced. But rather interestingly, after the three annealing steps, we observe that the room temperature wavelength of the mode is permanently shifted to the blue by approximately 0.5 nm. The maximum temperature used in this case is of approximately 130 °C.

We repeated the experiment at higher power in air and under vacuum. In both cases, the modes can be permanentely blue-shifted, up to 5 nm in air and 2 nm in vacuum, before vanishing of the signal. The quality factor of the modes remains fairly constant ($Q \simeq 900$) during the process, but the luminescence intensity drops steadily up to one order of magnitude. The highest temperature reached before damage was approximately 400-450 °C. The cause for this loss of photoluminescence is not fully understood yet, but could be due

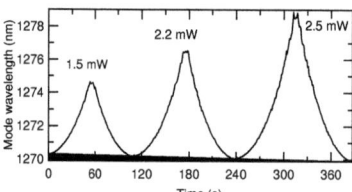 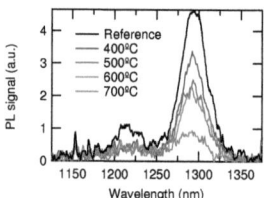

Figure 3.14: (left) Wavelength of a cavity mode as a function of time for three successive heating processes at 1.5 mW, 2.2 mW, and 2.5 mW. We observe that the initial energy of the mode blue shifts by approximately 0.5 nm during the whole experiment, as suggested by the black area below the curves. (right) Rapid thermal annealing performed on a similar sample capped with SiO$_2$ at four different temperatures.

to In migration or As desorption. However, this behavior is in good agreement with rapid thermal annealing measurements performed on a similar QD sample. Capping the sample with SiO$_2$ could reduce the damage on the luminescence, as displayed in figure 3.14b. The rapid thermal annealing was performed in a controlled oven during 30 seconds for 400 °C and 500 °C, and 60 seconds for 600 °C and 700 °C.

3.5 Global and local infiltration with polymers

In this section, we will examine a non-dynamical tuning strategy using the local infiltration of planar photonic crystals with UV-curable polymers. The work was performed in collaboration with Pascale El-Kallassi[f] who published a detailed study in[El-Kallassi 08]. The ultimate goal is to provide a fast and simple technique for producing optical circuits on a pristine photonic crystal template. Intonti *et al.* already showed the feasibility of this idea by using a relatively complex micro-infiltration system to fill single holes with a liquid [Intonti 06]. Here, we present a simpler way to achieve the same goal. The photonic crystal is immersed in a liquid monomer, and a focussed UV laser beam induces local polymerization. The remaining monomer is then washed away. The process is fast and relatively easy to implement on a large scale. We decided to test this method near a photonic crystal cavity, as the sensibility of the modes to modifications of the dielectric environment provides

---
[f]Group of Prof. Zuppiroli, Laboratoire d'optoélectronique des matériaux moléculaires, EPFL, Lausanne

3.5. Global and local infiltration with polymers

a precise tool to evaluate the success of the operation.

Infiltration technique

We studied L3 defect cavities, with a measured filling factor F of 34%, etched through a 320 nm membrane containing three layers of high density QDs at its center. The cavity shows three resonance peaks, with the middle one being a superposition of four modes with nearby energies. A graphical representation of the modes profile is provided on figure 3.15.

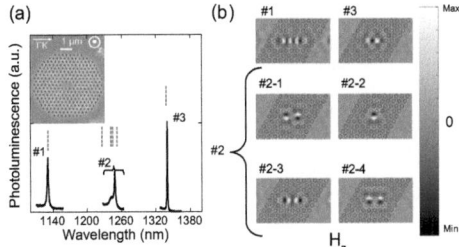

Figure 3.15: (a) Photoluminescence spectrum collected from an L3 cavity (inset SEM image: $a = 330$ nm and hole diameter $r_0 = 101$ nm). The red lines show the calculated resonance wavelength for a filling factor $F = 0.34$. (b) Calculated Hz-field maps of the TE cavity modes.

The photonic crystal structures are infiltrated with a liquid mixture of a dimethacrylate monomer (1,4-butanediol dimethacrylate) and 2% in weight of a photosensitive initiator (1-hydroxycyclohexyl) phenyl-ketone with broad absorption peaks in the wavelength region 220–375 nm. To ensure that the monomer penetrates in the holes of the photonic crystal, the sample was loaded into a vacuum chamber. After evacuation, the monomer mixture is poured over the sample, so that it is completely covered. Then, the chamber pressure is elevated at atmospheric conditions, forcing the monomer inside the holes. Simulations of the modes position and SEM observations indicate a monomer filling efficiency of $(95\pm5)\%$.

The photo-polymerization is performed with an Ar-ion laser emitting at 531 nm. This step is conducted under nitrogen flow to prevent the reaction of the free radicals with the ambient oxygen, as this would stop the reaction. The size of the laser spot, its intensity and position on the sample can be adjusted to perform large area polymerization as shown in figure 3.16. The residual monomer is washed away from the polymerized regions with organic solvents.

We observe a ∼20% shrinkage of the infiltrated material upon polymerization, and SEM measurements indicate that there is a residual layer of a few tens of nanometers of polymer covering the surface of the illuminated region.

Global and local polymerization

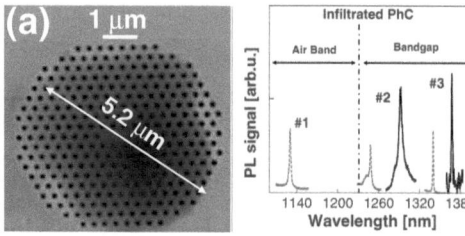

Figure 3.16: (Left) SEM top view of the L3 cavity and (right) PL spectrum showing the cavity modes (solid lines) and calculated position of the air band after global polymerization of the photonic crystal. The modes of the empty cavity are shown as reference (dashed lines).

The polymer fills the holes of the photonic crystal with a refractive index of 1.51 ± 0.01, higher than n_{air}. This results in a lowering of the air band energy, clearly visible for the case of global infiltration. The mode at higher energy of the unfilled cavity, labeled #1 in figure 3.16, do not appear after polymerization anymore, as its energy lays now outside the band-gap. The other modes red-shift by 45 nm for #2 and 28 nm for #3.

The Ar-ion laser can be focussed to a smaller spot size, in the order of 1 μm, to polymerize only a few tens of holes. In this case, it acts only as a local perturbation in the photonic crystal lattice: if close enough to the cavity, it will affect each mode differently, depending on their specific field profile. For example, in figure 3.17, the polymer is located near the long side of the L3 cavity, in the ΓM direction. The modes with profiles overlapping the infiltrated region are affected more substantially ($\Delta\lambda$ =6–10 nm) than the others ($\Delta\lambda$ =1.3 nm). The seemingly doubling of the high energy mode is indeed due to the emergence of a new cavity mode from the air band. Their respective field profile is displayed as an inset in figure 3.17 and the calculated resonance wavelength corresponds well to the

3.5. Global and local infiltration with polymers 67

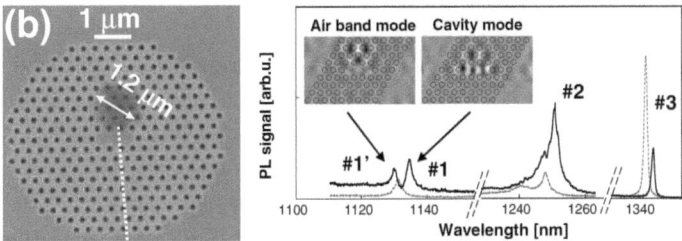

Figure 3.17: (Left) SEM top view of the L3 cavity and (right) PL spectrum showing the cavity modes (solid lines) after local polymerization in the ΓM direction. The modes of the empty cavity are shown as reference (dashed lines). Inset: field profile of the two modes around 1130 nm.

measured ones. The new mode is concentrated mainly in the polymerized region, that seems to act as a new cavity.

Local infiltration performed in the ΓK direction produces very similar results. Although, as the penetration of the field is smaller in this direction, the measured shifts were not as important, with $\Delta\lambda$ =0.1–3 nm depending on the mode.

Conclusions

The tuning of cavity modes by polymer infiltration shows large shifts, in the order of 30–45 nm for global and 1–10 nm for local infiltration. However, this method has some drawbacks for our application: the shift is not gradual, the quality factor of the modes shows a degradation in the order of 20% upon polymerization and the effective mode volume is enlarged, as the lowering of the refractive index contrast increases the penetration of the mode field into the holes. Still, this proves really interesting, for example in the writing of structures on a pristine photonic crystal sample [Intonti 06] or for other implementations where local infiltration is required [Mingaleev 04]. It also illustrates the fact that a perturbation near the cavity affects the modes differently, depending on the amount of overlap with their field profile. This last observation will be the subject of the next section.

3.6 SNOM spectral tuning

In this section, we report on a reversible spectral tuning of the resonances of a two-dimensional photonic crystal nanocavity induced by the introduction of a subwavelength size glass tip. The comparison between experimental near-field data, collected with $\lambda/6$ spatial resolution, and results of numerical calculations shows that the spectral shift induced by the tip is proportional to the local electric field intensity of the cavity mode, as mentioned in section 3.5. Incidentally, this observation proves that the electromagnetic local density of states in a nanocavity can be directly measured by mapping the tip-induced spectral shift with a scanning near-field optical microscope. The SNOM measurement were performed on our samples by Francesca Intonti at the LENS and department of physics of the University of Florence and the results published in [Intonti 08a]

Experimental method

A commercial SNOM (Twinsnom, OMICRON) is used in an illumination-collection geometry with a spectral resolution of 0.1 nm and spatial resolution around 200 nm. The sample is excited with a cw laser diode (780 nm) coupled into a pure dielectric, uncoated, and chemically etched near-field fiber tip raster scanned at a constant height smaller than 5 nm above the surface.

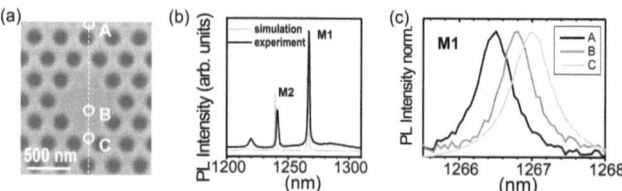

Figure 3.18: (a) Scanning electron microscopy (SEM) image of the investigated sample. The three white circles indicate the position of the tip where the spectra of (c) were collected. The dashed line indicates the sample region considered in figure 3.20. (b) Experimental (black line) and calculated (gray line) PL Spectra. (c) Near-field PL spectra collected at the three different tip position indicated in (a).

The measured structure consist of a D2 cavity, formed by four missing holes in a

3.6. SNOM spectral tuning

diamond-like geometry, in a triangular photonic crystal with filling factor $F = 35\%$ and lattice parameter $a = 301$ nm. It is etched through a 320 nm thick suspended GaAs membrane with three layers high density InAs QDs. Figure 3.18 displays a SEM image of the cavity, together with its PL spectrum.

Photoluminescence spectra

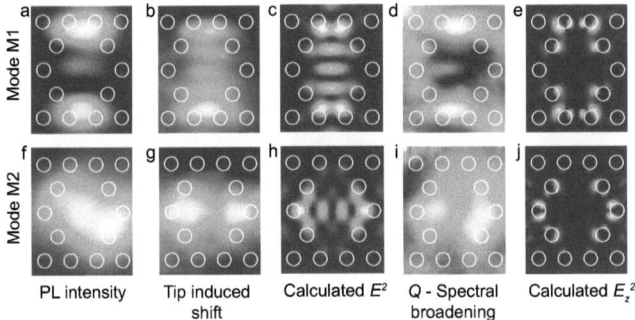

Figure 3.19: (a) Spatial distribution of the PL signal related to the resonance M1, integrated between the wavelengths 1266.2 nm and 1267.2 nm. (b) Map of the tip-induced spectral shift of the resonance M1. The maximum redshift is 0.5 nm. (c) Calculated spatial distribution of E^2 related to mode M1. (d) Map of the spectral broadening of resonance M1: minimum broadening $\Delta\lambda = 0.52$ nm ($Q = 2435$) and maximum broadening $\Delta\lambda = 0.65$ nm ($Q = 1950$). (e) Calculated spatial distribution of E_z^2 related to mode M1. (f) Spatial distribution of the PL signal related to the resonance M2, integrated between the wavelengths 1240.2 nm and 1241.2 nm. (g) Map of the tip-induced spectral shift of the resonance M2. The maximum redshift is 0.4 nm. (h) Calculated spatial distribution of E^2 related to mode M2. (i) Map of the spectral broadening of resonance M2: minimum broadening $\Delta\lambda = 0.56$ nm ($Q = 2215$) and maximum broadening $\Delta\lambda = 0.75$ nm ($Q = 1650$). (j) Calculated spatial distribution of E_z^2 related to mode M2.
The position of the photonic crystal pores is superposed to guide the eye. All the images cover an area of $1.15 \times 1.50\,\mu m^2$ and are plotted on a blue to white color scale.

The cavity shows two intense resonances M1 and M2, centered at 1266.8nm and 1241.3nm, with Q factors of 2400 and 2200 respectively, and orthogonal polarization. Interestingly, in the spectra collected through the fiber tip, the modes energy are clearly shifted, up to 0.5 nm when recorded at different points across the structure (figure 3.18a and 3.18c).

A complete 2D map of this spectral shift, produced by fitting the mode spectrum with a Lorentzian curve and extracting the central wavelength at each tip position, is displayed on figure 3.19b and 3.19g. The maximal red shift for M1 is of 0.5 nm and occurs at the vertical apexes of the diamond shape, while for M2 this occurs at the horizontal apexes and amounts to 0.4 nm.

Intuitively, the perturbation induced by the dielectric tip will be stronger where the mode profile is large. Inversely, at locations where the field is minimal or nonexistent, the frequency of the mode will not be perturbed by the fiber tip. Following theoretical predictions of a direct correspondence between the frequency shift and the unperturbed mode profile [Koenderink 05], we also plot the spatial distribution of E^2 on figure 3.19c and 3.19h.

Taking into account the somewhat limited resolution of the SNOM, the agreement between the map of the measured shift and the calculated E^2-field distribution is striking. It is in any case much more precise than the map of the PL intensity reported on figure 3.19a and 3.19f.

This observation has two consequences. On one side, that large shifts are obtained where the perturbation overlaps significantly with the field profile, as presumed in section 3.5. On the other side, that accurate recording of the field profile are obtained by mapping the tip induced shift in the mode frequency and not by directly mapping the PL intensity.

From the Lorentzian fits, it is also possible to extract the spectral width $\Delta\lambda$ of the perturbed modes (figure 3.19d and 3.19i) and thus gain information about the quality factor Q. The relation between the mode shift and Q is non-trivial, as the losses are expected to be higher not only for large field intensity, but also for tip positions above region with high E_z^2 (figure 3.19e and 3.19j). The spectral shift and mode width for a vertical scan across the cavity are represented in figure 3.20 for the mode M1. Although the position corresponding to the largest shifts also shows the strongest decrease in Q of roughly 20%, they are some regions, particularly for tip positions 0.2 nm away from the center of the cavity, where the Q factor has a global maximum for a relatively large mode shift. This is an exciting result, as it shows that it is possible to perturb the mode energy without altering its Q. Similar results have been obtained for the second mode M2.

3.7. SNOM local heating

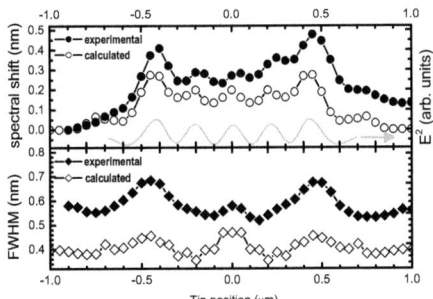

Figure 3.20: Upper panel: Experimental (filled circles) and calculated (open circles) tip-induced spectral shift of the M1 resonance. The gray line reports the vertical cross section, 200 nm wide, in the middle of the calculated electric field intensity map reported in figure 3.19c. Lower panel: Experimental (filled diamond) and calculated (open diamond) tip-induced spectral broadening of the M1 resonance.

Conclusion

Intonti et al. reported a new method to measure the LDOS with a SNOM by mapping the tip induced spectral shift of the mode frequency [Intonti 08a]. This method offers a better resolution that the usual mapping of the PL intensity [Intonti 08b].

They also demonstrated that the spectral tuning of the cavity mode through the local perturbation induced by a dielectric SNOM tip can produce mode red shift up to 0.5 nm, however with a degradation on the Q factor of approximately 20%. Still, for some locations, it is possible to achieve loss-less tuning of the cavity with a somewhat reduced amplitude around 0.3 nm. Next section will present a path to increase the amplitude of the this loss-less shift.

3.7 SNOM local heating

In this section, we present an extension of the SNOM tip induced shift method by exploiting the sub-wavelength local heating induced by intense near field laser excitation. The temperature gradient due to the optical absorption results in an index of refraction gradient which modifies locally the dielectric anatomy of the cavity and shifts the optical modes. Here, we show that reversible tuning can be obtained not only by changing the excitation power density but also by exciting at different locations in the nanocavity, with a resolution not accessible by laser induced heating as presented in section 3.4. The experiment was performed on our samples by Silvia Vignolini at the LENS and department of physics of the University of Florence and the results published in [Vignolini 08].

Experimental setup

The cavity is formed by four missing holes organized in a diamond like geometry (D2) in a triangular photonic crystal with lattice parameter $a = 301$ nm and filling factor $F = 35\%$ etched through a 320 nm thick membrane with three layers of high density QDs embedded in its middle. A scanning electron microscope image is reported in figure 3.21a.

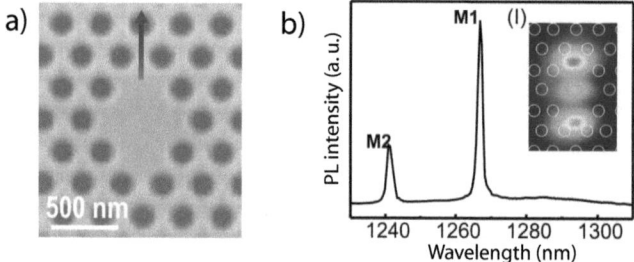

Figure 3.21: (a) Scanning electron microscope image of the investigated sample. The arrow corresponds to the scan from figure 3.22d. (b) Typical photoluminescence spectrum collected by near field probe at a power density of 14 MW/cm^2. The inset reports the spatial intensity distribution associated to the mode M1, the white circles superimposed on the image indicate the topographic positions of the photonic crystal holes. The two maxima of the spatial intensity distribution of M1 are located at the vertical edges of the cavity region.

A commercial scanning near field optical microscope (SNOM) (Twinsnom, OMICRON) is used in an illumination/collection geometry with a combined spatial and spectral resolution of 250 nm and 0.1 nm respectively. The lateral spatial resolution is directly extracted from the photoluminescence images and is defined as the full width at half maximum of the

3.7. SNOM local heating

profile of the smallest feature that can be resolved. The QDs are excited with 780 nm light from a CW diode laser coupled into a chemically etched, uncoated near-field fibre probe, that is raster scanned at a constant height above the sample surface. Photoluminescence (PL) spectra from the sample were collected at each tip position through the same probe and the PL, dispersed by a spectrometer, was detected by a cooled InGaAs array.

Measurements

Figure 3.21 shows a typical PL spectrum of the structure under investigation characterized by two main peaks, M1 and M2, centred around 1266.8 nm and 1241.3 nm, respectively. As both modes show a similar behavior, we will concentrate on M1 ($Q > 2400$). The inset of Figure 3.21b shows the $1.35 \times 2\,\mu m^2$ image of the integrated PL intensity in the wavelength range between 1265 nm and 1267 nm (i.e. integrated over the optical mode). This PL intensity map is predominantly concentrated at the diamond tips with a lobe of lesser intensity in the centre of the structure. It reproduces qualitatively the spatial distribution of the electric field associated to the resonance [Intonti 08b].

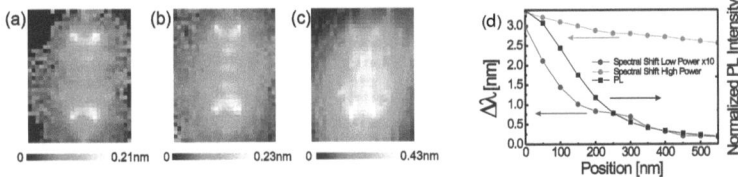

Figure 3.22: (a), (b) and (c) Maps of the tip induced spectral shift (see section 3.6) for different low excitation power densities: $0.3\,MW/cm^2$, $0.7\,MW/cm^2$ and $1.4\,MW/cm^2$. (d) Spectral shift of the mode for a tip position scanned from the top apex of the cavity until 500 nm north of it, i.e. along the arrow in figure 3.21a. The PL intensity is reported on the right axis for comparison.

In figure 3.22a–c, we see the map of the tip induced spectral shift (see section 3.6) at three different low excitation power densities. At low powers, it reproduces the square modulus of the electric field E^2 of the mode [Intonti 08a]. However, as the intensity of the excitation is increased, this induces a broadening of the measured shift.

Figure 3.22d displays the magnitude of the shift for a scan starting at the maximum of the mode field, and extending away from it in the ΓM direction. The path is illustrated by an arrow on figure 3.21a. These spatially decaying curves give us an idea on the refractive index gradient, and thus on the laser induced heat gradient, for different excitations. We notice that for low excitation, the tip induced shift decays spatially at a similar rate as the PL intensity and the heating effect can be neglected. For high excitation, the spectral shift barely decay over the 500 nm range, and can thus be approximated by a global temperature heating, at least in the spatially non-resonant case.

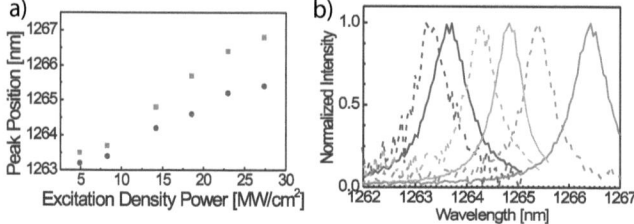

Figure 3.23: (a) Central wavelength of the mode as a function of the excitation power. The red squares correspond to a spatially resonant excitation (i.e. with the tip above a maximum of the mode electric field) on the apex of the diamond shape (see figure 3.21) and the blue dots are collected with the tip spatially off resonance, 0.5 μm north of the apex. (b) Near field spectra obtained for three different power excitation densities (5 MW/cm^2, 14 MW/cm^2 and 23 MW/cm^2 in blue green and red, respectively) of the mode M1. Continuous line spectra are on spatial resonance, whereas dotted ones are off resonance.

If we look at the mode wavelength for spatially resonant excitation (i.e. the spot is located at the beginning of the arrow on figure 3.21a) and the spatially non resonant case (the spot is at the end of the arrow), we see that they do not shift at the same speed for increasing power excitation (figure 3.23). Vignolini *et al.* show that the non resonant case can be well explained by a global heating of the whole structure of approximately 18 K [Vignolini 08].

However, in the resonant case, the heat source is located directly at the maximum of the mode field and cannot be treated as constant anymore: in a similar way as an external dielectric perturbation has a stronger influence at field maximum, the modification of the refractive index, due to the heat source, have the largest effect on resonance. Even if the lateral decay is small, the fact that the mode field is maximal amplifies the effect of the refractive index gradient. Vignolini et al. used FDTD simulations to reproduce their results numerically. In addition of the 18 K global heating, they assumed a localized heat source, with the same size as their SNOM tip. They simulated it by changing the refractive index on a cylindrical portion of the GaAs membrane until they reproduced the shift observed experimentally. The best fits were for cylinder diameters between 300 nm and 400 nm with temperatures increase by 100 K to 75 K respectively, which confirms the local nature of the heating.

Conclusion

In conclusion we successfully demonstrated a dynamic tuning of a two dimensional photonic crystal nanocavity by spatially resonant (3.5 nm red shift) and non-resonant (2.1 nm red shift) heating of the sample. This technique has the advantage of being completely reversible and allows to continuously tune the cavity modes with no substantial degradation in Q and no significant modifications of the field distribution associated to the mode. This could be used for example in coupled cavity devices [Vignolini 09], where the supermode is tuned through one cavity, and the QDs are left unperturbed in the other cavity.

3.8 Double membrane tuning

An electromagnetic wave is sensible to modifications of the dielectric environment in which it propagates. For example, by approaching a dielectric object close to the cavity, we can perturb the propagation of the wave, and thus modify the energy of resonant modes, as mentioned in section 3.6. Instead of using an external perturbation, as an AFM [Märki 06, Hopman 06] or a SNOM probe [Lalouat 07, Mujumdar 07, Intonti 08a], we could imagine to build a cavity with a tunable geometry. A modification of its configuration will then affect directly its mode distribution.

Based on the idea first proposed by [Notomi 06], we designed a double-membrane structure with a tunable inter-membrane distance, as depicted in figure 3.24. Each membrane has a thickness of 165 nm, half that of the usual samples and the quantum dots are located at the center of the top membrane. The photonic crystal structure is etched through both membranes.

Intuitively, a mode guided in the top membrane will have an evanescent tail overlapping

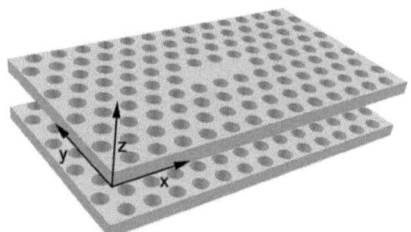

Figure 3.24: Sketch of the double membrane device. The holes are drilled all the way through the structure. We introduce a xyz coordinate system with origin on the inner surface of the top membrane and with the z direction perpendicular to the membranes surface.

with the bottom membrane. By changing the distance between the membranes, we can control the amount of dielectric material perceived by the mode. This, in turn, will affect its effective refractive index, and its interaction with the photonic crystal.

Formally, we can decompose the three-dimensional confinement of the cavity in an in-plane component due to the photonic band-gap, and an out-of-plane component due to the wave-guiding properties of the membrane. In other words: $F(x,y,z) = F_{\text{PhC}}(x,y) \cdot F_{\text{WG}}(z)$. If the perturbation acts uniformly on the photonic crystal, i.e. $\varepsilon(x,y,z) = \varepsilon_o + \delta\varepsilon(z)$, we can assume that it will affect mainly the wave-guided modes and modify their effective refractive index $n_{\text{eff}} = \frac{c_o}{v_g}$ without perturbing the photonic band-gap properties.

Notomi *et al.* used the three dimensional finite-difference time domain (3D-FDTD) method to compute the double-membrane cavity structure [Notomi 06]. They observed that when the distance between the membranes is changed, the quality factor Q and the effective mode volume V_{eff} remain fairly constant. Though surprising, they explain this by the fact that the inter-mambrane distance Δ does not significantly affect the in-plane k-space distribution, which mainly determines the out-of-plane leakage [Srinivasan 02, Akahane 03]. They also confirmed that the changes in the resonant wavelength of the cavity modes are mostly explained by the changes in the effective index n_{eff} of the wave-guided modes as a function of Δ.

Modes of the double-membrane waveguide

We used our usual plane-wave-expansion (PWE) approach to calculate the cavity modes of the structure. We started by analyzing the double membrane waveguide.

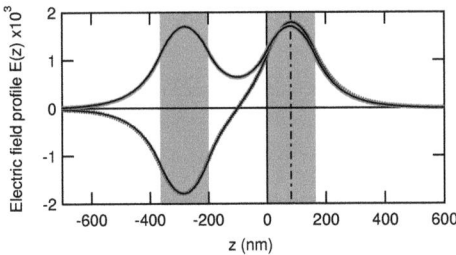

Figure 3.25: Symmetric (red) and anti-symmetric (blue) guided modes of the double membrane structure for a 200 nm wide separation. The gray areas represent the GaAs membrane, and the vertical dashed-doted line indicates the position of the QDs. The thick curves are the result of a fully vectorial 2D mode solver, whereas the thin lines are computed as the sum and difference of the single membrane wave-guided mode. This approximation holds for membrane separations larger than 50 nm.

Intuitively, this system can be seen as two independent single-mode planar waveguides W_1 and W_2 far from each other. As the distance Δ decreases, the overlap between W_1 and the evanescent field of the mode in W_2 increases. From the superposition principle, we get coupling between the modes in W_1 and W_2. This coupling induces a splitting in a symmetric and anti-symmetric mode (see figure 3.25) with different energies $\omega^{as} > \omega^s \implies n^s_{\text{eff}} > n^{as}_{\text{eff}}$, similar to the treatment of the H_2^+ molecule in quantum mechanics.

We thus obtain two different effective indexes starting from the values of the two first modes $n^{(1)}_{\text{eff}}$ and $n^{(2)}_{\text{eff}}$ of a 330 nm thick waveguide for $\Delta = 0$, and merging to the effective index n^∞_{eff} of the 165 nm thick membrane for $\Delta \to \infty$. They vary exponentially with the distance, with the same decay constant q as the evanescent field of the thin membrane.[g]

$$n_{\text{eff}}^{s} = n_{\text{eff}}^{\infty} + \left(n_{\text{eff}}^{(1)} - n_{\text{eff}}^{\infty}\right) e^{-q \cdot \Delta} \qquad (3.1a)$$

$$n_{\text{eff}}^{as} = n_{\text{eff}}^{\infty} + \left(n_{\text{eff}}^{(2)} - n_{\text{eff}}^{\infty}\right) e^{-q \cdot \Delta} \qquad (3.1b)$$

where

$$q = \frac{2\pi}{\lambda_0} \sqrt{\left(n_{\text{eff}}^{\infty}\right)^2 - n_{\text{air}}^2}$$

Figure 3.26 displays the symmetric and anti-symmetric effective refractive index of the double-membrane structure as a function of the membrane separation. The refractive index of GaAs at room temperature, with $\lambda_0 = 1300$ nm, is $n_{GaAs} = 3.41$ [Marple 64]. Solving the modal equation[h] yields $n_{\text{eff}}^{\infty} = 2.66$, $n_{\text{eff}}^{(1)} = 3.11$, and $n_{\text{eff}}^{(2)} = 2.08$.

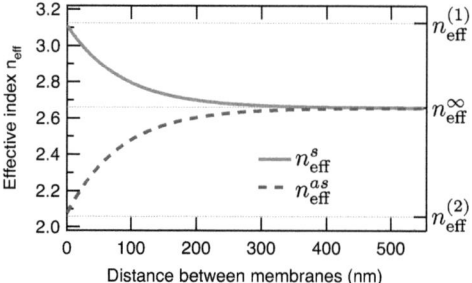

Figure 3.26: Effective refractive index n_{eff} computed from equation 3.1 for two 165 nm wide membranes as a function of their separation. The material refractive index of GaAs, $n_{\text{GaAs}} = 3.41$, was taken from [Marple 64] for $\lambda_0 = 1300$ nm at room temperature, and $n_{\text{eff}}^{\infty} = 2.66$, $n_{\text{eff}}^{(1)} = 3.11$, and $n_{\text{eff}}^{(2)} = 2.08$.

Effect on the modes energy of L3 and H1 cavities

Since the two waveguided modes will be affected differently by the photonic crystal, we need to calculate the dispersion relation of the structure. The air-band edge, the single mode

---
[g]This result was confirmed by simulations with a fully vectorial 2D mode solver, MODE Solutions 2.0, from Lumerical

[h]see for example [Saleh 91] on chapter 7 or [Rosencher 02] on chapter 9

3.8. Double membrane tuning

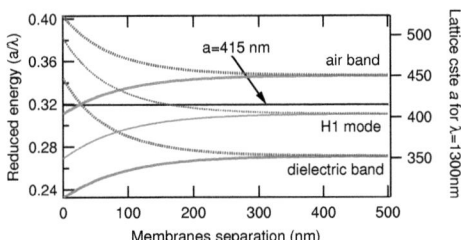

Figure 3.27: Photonic band-edges for the symmetric (thick solid red line) and anti-symmetric (thick blue dashed-line) modes as a function of the membrane separation for a photon wavelength of 1300 nm. The thin lines represent the mode of a H1 cavity. The right axis indicates the corresponding lattice parameter. A QD emitting at $\lambda=1300$ nm in a structure with lattice parameter $a=415$ nm (horizontal line) will couple to the anti-symmetric H1 mode for a membrane separation of $\Delta \simeq 160$ nm.

of a H1 cavity and the dielectric band edge energies are represented as a function of the inter-membrane distance Δ on figure 3.27. They were computed with the PWE method, starting with the material refractive index of GaAs for 1300 nm light. As expected, the energies for the symmetric and antisymmetric waveguided modes merge for large Δ. We notice that they start diverging significantly (>1%) as Δ becomes smaller than 250 nm, and that the two bandgaps even get fully disconnected for $\Delta < 28$ nm.

The right axis of figure 3.27 gives the photonic crystal lattice parameter a corresponding to the wavelength 1300 nm. If we constructed a structure with $a = 415$ nm (horizontal line), and observed a single QD emitting at 1300 nm we would witness some very interesting behavior. As the inter-membrane distance is reduced, the anti-symmetric mode will draw nearer to the QD from the red side, until they reach resonance for $\Delta \simeq 160$ nm. Then the mode will continue drifting away to the blue side. Nothing really new as it seem. But wait: when Δ reaches below 28 nm, the QD will leave both the symmetric and anti-symmetric bandgaps through the air-band edge, respectively the dielectric band edge. This would allow one to measure the dynamic of the same single QD not only in resonance and out of resonance with a cavity mode, but also out of the cavity, since there is no cavity effect anymore out of the bandgap. Moreover, this last measurement would include any dynamic behavior disturbance caused by eventual defect states attributable to the proximity of the photonic crystal holes surface.

It would be tempting to calculate the amplitude of the mode shift by using the reduced energy values given in figure 3.27. However, we have to take into account the material dispersion to have a realistic idea of the achievable tuning (see Appendix A)[i].

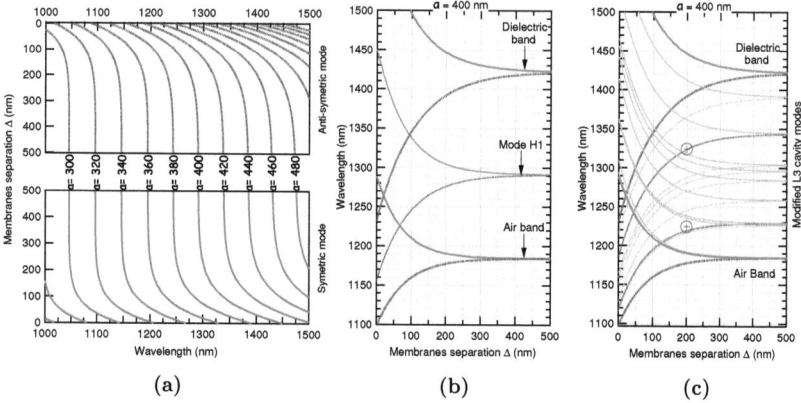

(a) (b) (c)

Figure 3.28: (a) Lattice constant value a (written vertically between the plots) of the single mode of a H1 cavity for the anti-symmetric (top half) and symmetric (bottom half) waveguided mode as a function of the membrane separation Δ. The emission wavelength of the mode for a given lattice constant a can be read on the bottom axis. (b) Photonic band edges and H1 cavity mode wavelength as a function of Δ for $a = 400$ nm. The blue dotted lines are for the anti-symmetric waveguided mode. (c) Photonic band-gap and modified L3 cavity modes wavelength as a function of Δ for $a = 400$ nm. The points at 1224.4 nm and 1324.4 nm for $\Delta = 200$ nm correspond to the measured wavelengths of the two modes studied in figure 3.40 on page 91.

Figure 3.28a is a two-dimensional map of lattice parameters a matching the energy of the H1 mode as a function of the inter-membrane distance (left axis) and light wavelength (bottom axis). As it takes into account the material dispersion of GaAs, it is a really helpful tool in estimating visually the emission wavelength of the mode: one only needs to follow the

---

[i]If we used the reduced energy values for the H1 cavity mode to calculate the mode tuning range while starting at 1300 nm for $\Delta \to \infty$, we would obtain a 244 nm blue shift and a 198 nm red shift for $\Delta = 0$.

3.8. Double membrane tuning

lines of constant lattice parameter. The upper part of figure 3.28a is for the antisymmetric mode. It has its left axis inverted to allow a smooth transition from the symmetric mode on the lower half of the figure. The lattice parameter a is written vertically between the two graphs. If we follow the value corresponding to a mode at 1300 nm for large Δ, we obtain a 161.3 nm red shift for the symmetric mode and 138.5 nm blue shift for the antisymmetric one when the membranes are in close contact. When compared to the linewidth of a cavity with even a moderately low quality factor Q, and to the tuning range of other methods, one realizes the incredible potential of effective index tuning as presented here. Figure 3.28b and 3.28c display the wavelength of a H1 cavity and a modified L3 cavity respectively, as a function of the inter-membrane distance for a lattice parameter a=400 nm. These figures take into account the GaAs material dispersion. As the left scale is in nanometers, the structures seem inverted from their position in figure 3.27, and the dielectric band is displayed above the air band. The modified L3 cavity in 3.28c has first hole diameters shifted outwards by 15% and reduced by 2/3. We fixed the lattice parameter a to 400 nm, has it is the same as the cavity DM3_L3_400s measured below (see figure 3.40 on page 91). The position of two modes observed in microPL are reported on the graph.

Measurement setup

(a) Picture of the setup   (b) CCD image of the needle

Figure 3.29: (a) Setup used to apply pressure on the top membrane while recording the cavity mode energy shift. The optical path is the same as in usual confocal configuration. Here, a STM tip, fitted in the tube of a hypodermic needle, can be positioned precisely with one of the xyz-piezo driven stage. See figure 3.33 right for a close-view. (b) The tip of a hypodermic needle in contact with the sample surface.

To measure the effects of the inter-membrane distance on the mode energy, we had to find a way to apply small displacements (several nanometers) on the top membrane. We needed a tool with a tip size smaller than a few tens of micrometer, but long enough to be manipulated between the sample and the microscope objective. At first, we used STM probes fabricated by our co-workers Joris Keizer and Cem Çelebi at TU/e. We attached the probes to one of the xyz-piezo driven stage of the TRIAXsetup to obtain accurate positioning on the sample, as illustrated on figure 3.29.

However, due to their tenuity, the STM probes have a tendency to bend permanently and thus slide over the surface of the sample rather than applying pressure on it. We then turned to hypodermic needles: they have a really sharp apex, but the tubular structure give them enough rigidity to effectively bend the membrane. Still, mainly because they press on the sample at an angle, even the needles show a certain elasticity. It is then really difficult to estimate the amount of vertical displacement of the membrane.

As we were concerned that the needle or the STM tips could perturb the mode energy, we always positioned them away from the cavity center, usually at the border of the photonic crystal region. We compared spectra with and without needle, and couldn't measure any significant effect.

The approach of a 4 cm long needle to the border of a 20 μm wide photonic crystal region is quite delicate and can result in the crashing of several cavities if not done properly. However, there is a simple procedure to manage this without damage.

1. Turn on the CCD imaging and the green illumination LED. Select a cavity in the center of the image.

2. Visually approach the needle up to one millimeter of the sample surface.

3. Visually align the tip of the needle to the focalized LED light.

4. The CCD image is not necessarily fully dark, and even when the probe is in the optical path, it is possible to observe structures below(see figure 3.30). Move the probe in the $x-y$ plane of the CCD image until the its side is visible (dark shadow) and try to bring its tip in the center of the image

5. Carefully move the probe closer to the sample.

6. Repeat steps 4 and 5 until the tip of the probe is sharp, as in figure 3.29b.

3.8. Double membrane tuning

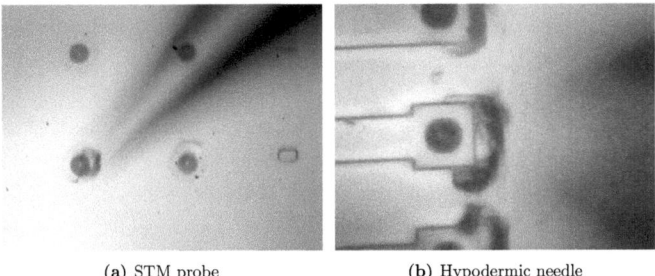

(a) STM probe (b) Hypodermic needle

Figure 3.30: Image illustrating the effect of a STM tip (a) and of a hypodermic needle (b) lifted above the sample surface (i.e. out of the focal plane of the CCD image)

Sample design

We produced two samples by MBE following the usual growth procedure with the 1.5 μm thick $Al_{0.7}Ga_{0.3}As$ sacrificial layer on top of the GaAs substrate. However, the membrane structure deposited atop consists of two 165 nm GaAs layers separated by a 100 nm (200 nm) $Al_{0.7}Ga_{0.3}As$ spacer for sample P938 (P920 respectively). The lower membrane is passive, whereas the upper one contains a single layer of high density InAs QDs emitting at 1300 nm.

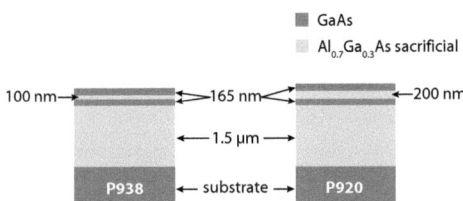

Figure 3.31: Structure of the samples P920 and P938 grown by molecular beam epitaxy (MBE) with a double membrane structure.)

Sample DM1

The first double membrane cavities were etched on the P920 sample with 200 nm inter-membrane distance. We used a mask with modified H1 and L3 cavities without any specific structure around them. The photonic crystal lattice parameters a were not adjusted to the double-membrane effective refractive index. In consequence, the fundamental mode is not centered around 1300 nm. Figure 3.32 shows some SEM pictures of the structure. We notice that the side-walls of the holes are pretty rough, that the holes in the second membrane are badly shaped, and that there is a structure dangling below each hole. This is most probably Al_2O_3 formed by oxidation of the AlGaAs between the RIE and under-etching steps, as also observed in [Braive 08] (page 72). Although we didn't expect high quality

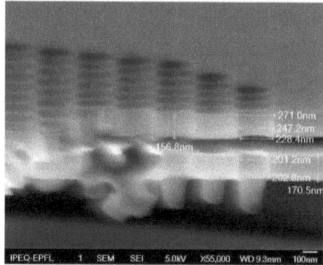 

Figure 3.32: SEM performed on sample P920 with modified H1 and L3 cavities. The side view displays the measured diameter of the holes at various points. They are rather conical in shape and an prolonged by a dangling structure, most probably Al_2O_3.

factors from this batch, we nevertheless decided to perform microPL measurements while pushing the membrane with a STM probe as illustrated on figure 3.33. At first, when the tip is on contact, we do not observe any wavelength shift. By further pressing, we cause a rupture in the membrane, clearly visible on the CCD pictures. We then start to observe a mode redshift reaching approximately 1.5 nm at is largest, which is in the same range as what we obtain with gas deposition at cryogenic temperature (see figure 3.10 on page 59). However, this is still quite far from the tuning range expected for double-membrane systems. As we further push the STM probe, it bends and slides on the sample surface. When the probe is lifted up, the mode goes back to its initial energy.

Sample DM3

We decided to try another design in order to free the membrane around the cavity. We designed cantilevers with different arm widths, and two types of spiral-shaped structures (figure 3.34). The latter would allow the top membrane to move vertically, since we were

3.8. Double membrane tuning

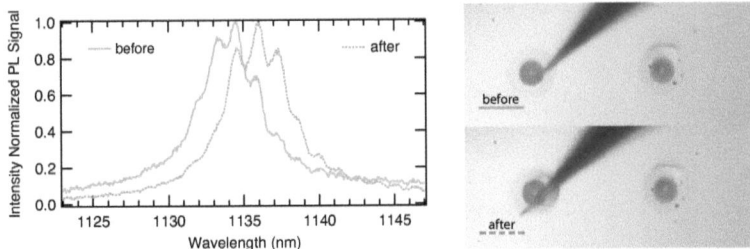

Figure 3.33: (left) Spectrum of a cavity mode before (sold line) and after (dotted line) pressing the membrane down with a STM probe. (right) The corresponding CCD images show how the probe bends the membrane by breaking it at one side (bottom image), and how it then tends to slide on the sample surface.

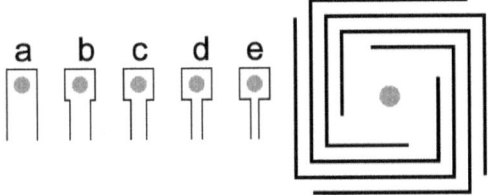

Figure 3.34: Some of the structures designed on the sample. The cantilevers have widths $(1.00 \times 30 \cdot a) = 30 \cdot a$, $(0.66 \times 30 \cdot a) = 19.8 \cdot a$, $(0.50 \times 30 \cdot a) = 15 \cdot a$, $(0.33 \times 30 \cdot a) = 9.9 \cdot a$, and $(0.25 \times 30 \cdot a) = 7.5 \cdot a$ for a, b, c, d, and e respectively. The arm length is $(1.33 \times 30 \cdot a) = 39.9 \cdot a$, where a is the photonic crystal lattice parameter.

concerned that in a single arm cantilever, the top cavity would shift off axis during the bending. We also adjusted the lattice parameter a to have the QD's emission centered inside the photonic bandgap. We used sample P920 again, with 200 nm inter-membrane width. Unfortunately, as seen on figure 3.35, fully under-etched structures have the tendency to collapse to the substrate. Moreover, as the etching rate of the inter-membrane layer is slower than for the sacrificial layer, we did not obtain totally free standing top membranes. We once more observe holes with rugged sidewalls and, in the lower membrane, with ill-formed shapes. We nevertheless carried on with microPL measurements.

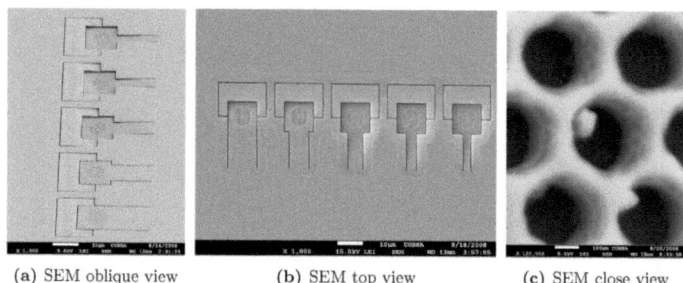

(a) SEM oblique view (b) SEM top view (c) SEM close view

Figure 3.35: (a) The angled view displays the bending of the cantilever arm when fully under-etched. (b) The top view gives an idea of the extent of the sacrificial layer removal after 180 s etching in 5% HF. (c) The close view of the photonic crystal holes reveals corrugated side-walls and rugged shapes.

An interesting example is cavity DM3_L3_370Cb, a L3 cavity with lattice parameter $a = 370$ nm on a cantilver with a relatively wide arm of 7.3 µm (type b in figure 3.34). The cavity exhibits four modes (m_1 to m_4) at 1219.0 nm, 1231.2 nm, 1316.8 nm, and 1344.7 nm before beginning with the tuning. We then pushed the membrane down with the STM probe. We measured blue shifts of -0.10 nm, -0.23 nm, -0.30 nm, and -0.29 nm for modes m_1 to m_4. We lifted the probe and measured the position of the mode again. Against all expectations, the mode had blue shifted even more, reaching -0.22 nm, -0.31 nm, -0.35 nm, and -0.37 nm.

We performed the experiment again after replacing the STM probe with a surgical needle. In the following we will concentrate on the second (and third) modes. The results for mode m_3 are summarized in figure 3.36.

It is noticeable that the mode blueshifted even further between steps C and 0, during the time when the STM-probe was replaced with the needle. As we start pushing again (step 1), we observe at first a 0.42 nm (0.25 nm) blue shift. Then, at step 4, the modes start redshifting. At the same time, we observed a bending of the cantilever on the CCD camera. Further pressing on the cantilever increases the shift towards longer wavelength, with a maximum amplitude of 1.5 nm (1.2 nm) for the fourth (third) mode. Interestingly, the whole process does not lead to a degradation in the cavity quality; we even observed a

3.8. Double membrane tuning

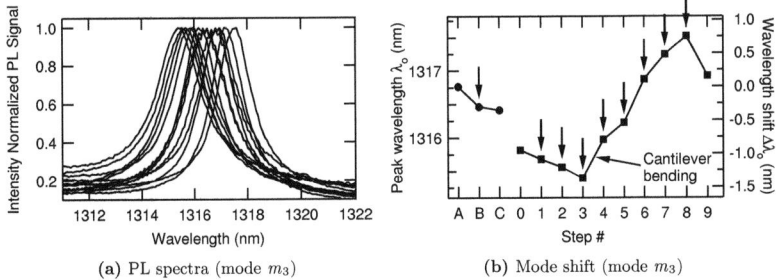

(a) PL spectra (mode m_3)

(b) Mode shift (mode m_3)

Figure 3.36: Shift of the cavity mode while bending the double membrane cavity DM3_L3_370Cb. (a) Normalized micro-photoluminescence spectra. (b) Mode wavelength obtained by fitting the cavity mode spectra from (a) with a lorentzian lineshape. Steps #A→#C are for the first experiment, and #0→#9 for the second. The vertical arrows (↓) indicate the data points measured while the top membrane was pushed down. Between points #3 and #4, we observed a bending of the cantilever on the CCD camera.

30% increase of the Q factor up to 850 for mode m_3. The last measurement (#9) was taken while the cantilever is relaxing back to the original state after the point was lifted off the sample.

This experiment leads to two puzzling observations: the mode seems to blue shift from itself at the beginning, even without applying pressure anymore, and the cavity modes changed the shifting direction from blue to red while applying pressure increasingly. However, as we recall that this reversal point occurred precisely at the time when we could see the cantilever bending leads to an tentative elucidation of this mystery. But let us first look at the effect of bending on a single membrane.

Single membrane bending

We performed this experiment on a sample with modified L3 cavities in a 320 nm thick membrane. We analyzed the effect of applying pressure on cavity P698_L3_330_e_x and plotted a close view at the modes centered around 1285 nm on figure 3.37.

The spectrum of the unperturbed cavity corresponds to measurement step #0. Surprisingly, we observe a clear redshift of 1.3 nm over all the experiment. The shift mostly

Chapter 3. Tuning

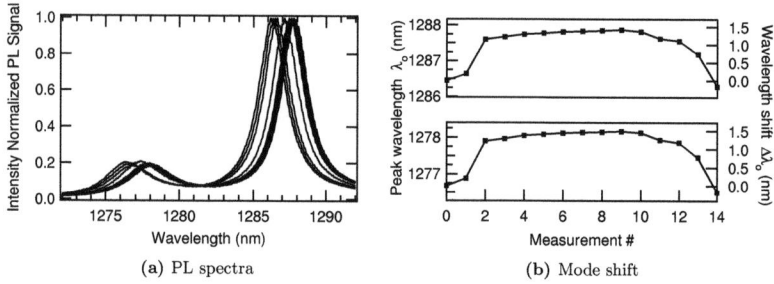

(a) PL spectra (b) Mode shift

Figure 3.37: Redshift of the cavity mode while bending a single membrane.

occurred between measurement points #1 and #2 (∼1.0 nm), between which the needle was actually not moved! The first push (step #1) was probably strong enough to induce a large tension in the membrane and in the needle. We waited 10 minutes before recording point #2, the membrane had time to bend significantly during this time. As we continued pushing, the wavelength continued to red shift, but at a slower rate. Finally, we lifted the needle at step #10. Step #11 was recorded after 5 minutes, step #12 after 12 minutes, step #13 after 20 minutes, and finally step #14 after almost 14 hours. We observe a total recovery in the cavity spectrum of the unperturbed membrane.

Before trying to explain the possible origin of this redshift on a single membrane, we performed some tests measurement to rule out any effect that could have been caused by misalignment. As a matter of fact, since the needle reaches the sample surface under an angle, pushing it perpendicularly towards the sample causes the needle to bend. This, in turn, makes it slide across the membrane surface, and in some case, slightly pushes the sample laterally. In addition, the membrane bending somewhat tilts the cavity with respect to the collection optic. It also moves it lightly out of focus. We usually correct for sample motion by realigning the laser spot above the membrane, maximizing the collected intensity. However, to rule out any of this effects as causes for the mode shift, we performed four series of measurements summarized on figure 3.38.

As a result, we obtained shifts smaller than 0.02 nm for scanning along the ΓK direction and perpendicularly to it across the whole cavity, smaller that 0.05 nm while zooming in and out of focus (500 μ-steps range), and smaller that 0.06 nm for tilting the sample up to 20°

3.8. Double membrane tuning

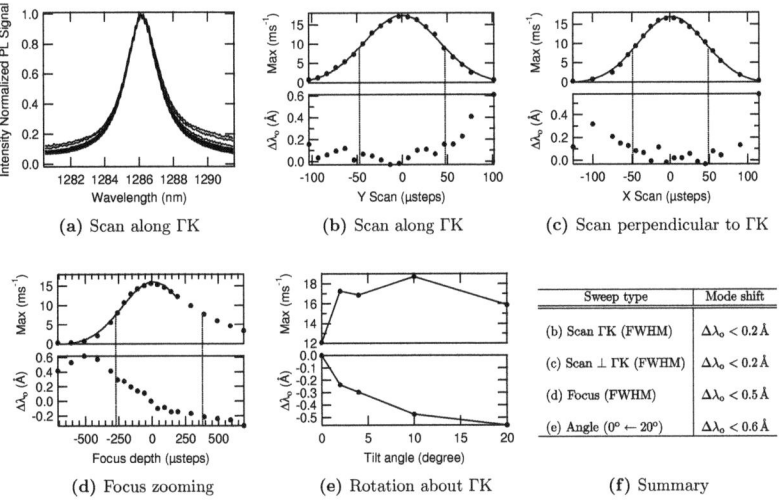

Figure 3.38: Effects of misalignment on the mode wavelength of a (modified L3) cavity. (a) Spectra corresponding to a scan along the ΓK direction (see (b)) illustrating the small extent of the wavelength shift. (b—e) For each position, the mode spectrum was fitted with a Lorentzian lineshape to extract its amplitude (top graphs) and central wavelength (bottom graphs). One μstep corresponds to a linear displacement of the piezoelectrical actuator of approximately 10 nm. (f) As we usually maximize the alignment before measuring, we considered the mode shift only between the FWHM boundaries (dotted vertical lines in b—d)

around an axis across the ΓK direction. As a side note, we observed a competition between two modes around 1230 nm while scanning along ΓK. Knowing the mode field pattern (see for example Appendix B on page 127), this could be used to identify them accurately.

They are basically two possible ways to explain the 1.3 nm redshift of the single membrane sample. Either through a structural deformation of the photonic crystal, or through a modification of the refractive index, including material refractive index or effective refractive index of the wave-guided mode.

Family picture of the double membrane

The experiments with the single membrane helped us to understand the origin of the mode shift reversal observed on figure 3.36: we are actually bending the whole double-membrane structure towards the GaAs substrate. This is in agreement with the observed bending on the CCD image between step #3 and #4. However, we still have to understand why the mode was blueshifting even without applying further pressure on the top membrane.

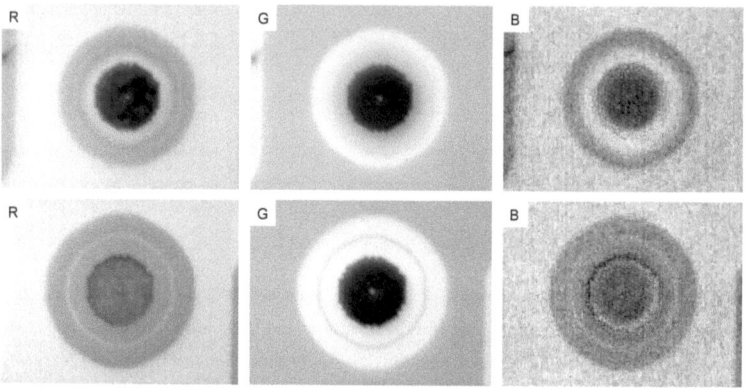

(a) RGB decomposed white light microscopy

(b) SEM images

Figure 3.39: (a) White light optical microscopy image of cavity DM3_H1_390SS with a collapsed top membrane (top row) and of DM3_H1_400SS with flat membranes (bottom row), decomposed in its Red, Green, and Blue components. (b) SEM cross section images illustrating the cases when the top membrane is collapsed (left) or not (right). The SEM images were recorded on sample DM1.

3.8. Double membrane tuning

If we look at the sample under white light optical microscopy, we can spot two different types of cavities, as illustrated in figure 3.39a. We decomposed the color images in their RGB (Red, Green, Blue) component to enhance the effect. We clearly recognize the cavity surrounded by a halo corresponding to the under-etched membrane. For the first type of cavity (first row in the figure), which we encounter approximately 95% of the time, the halo shows an intensity gradient form the cavity side until approximately half the distance to its border. We interpret this as a collapse of the top membrane, as illustrated in the left cross-sectional SEM image of figure 3.39b. The second type of cavity show a constant contrast across the whole halo. This indicates that the two membranes are equidistant, and thus that the top membrane is not collapsing. We also clearly see a circle in the middle of the halo, which corresponds most probably to the border of the inter-membrane under-etched region.

We measured cavity DM3_L3_400s, which has no gradient in its halo as can be seen on figure 3.40, and we observed a clear 3.3 nm (3.5 nm) blueshift of the mode at 1224.4 nm (1324.4 nm). From figure 3.28c, it is expected that the lower energy mode shifts faster than the other for the same displacement Δ.

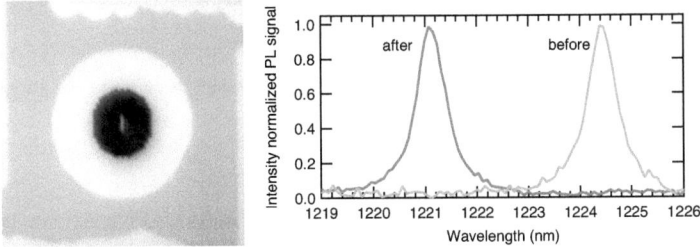

Figure 3.40: (left) White light microscopy and (right) microPL spectra of cavity DM3_L3_400s displaying a blue shift larger than 3.3 nm. The solid line is the cavity in its idle configuration, while the dotted line is for the case where the membrane is pushed down.

We have now all the keys at hand to explain satisfactorily the shifting behavior reported in figure 3.36 for cavity DM3_L3_370Cb. From step #B to #3, we approached the top membrane to the bottom one, thus causing a blueshift of the mode. Then, as we pushed further, the whole double-membrane structure started bending, which caused the observed redshift. However, there is still one feature that we didn't explain yet: why did the mode continue its blueshift on #C and #0, whereas nothing was pushing the membrane down?

Membrane sticking

The maximum dimensions of suspended multi-layered structures depend on the final drying procedure. During the elimination of the last rinsing liquid, capillary forces tend to collapse the released parts together. Strong adhesion forces, referred to as stiction forces in micromechanics, can then cause devices to remain permanently stuck. These surface forces are much stronger in the sub-micrometer regime, due to the increased surface to mass ratio, as compared to our familiar macroscopic environment.

This sticking could explain the self-induced blue shift of the mode. At point #2, we pressed with the STM probe at the border of the photonic crystal. We probably pushed so much, that we brought the two membranes in contact at that position. Then, even though we lifted the probe, the membrane remained stuck together due to stiction forces. The sticking region then diffused gradually across the photonic crystal, bringing the two membrane progressively closer at the position of the cavity, thus provoking a self-induced blueshift of the mode.

Critical point drying

Apparently, rinsing the sample in hot isopropanol is not sufficient to avoid the sticking of the membranes during the fabrication process. Among the existing drying methods (freeze drying, solvent evaporation, dry under-etching, etc.), CO_2 supercritical point drying has given the best results, according to [Garrigues 02].

We decided to try this approach and designed cavities on sample P938 with the 100 nm inter-membrane distance. After the HF wet etching, we rinsed the samples in water and kept them in ethanol overnight. We then asked to Antonio Mucciolo[j] to perform the critical point drying, using the same method as for biological samples. We removed the samples from their transport container and placed them in the high pressure chamber half filled with cold ethanol (6°C). We then completed the filling of the chamber with liquid CO_2 ($\sim$ 60 bar), waited 10 minutes and emptied half of the liquid. This process was repeated four times to replace the ethanol with liquid carbon dioxide. Then, the pressure was raised to 81 bar and the temperature to 42°C, which is above the critical point of CO_2 ($T_c = 31.1$°C and $P_c = 73.9$ bar). After that, the pressure was reduced to atmospheric condition in order

---

[j]Centre de microscopie électronique (CME), Faculté de biologie et médecine de l'Université de Lausanne

3.8. Double membrane tuning 93

to pass around the critical point without passing through the liquid/gas phase transition. Since a liquid/vapor interface is not formed, no capillary forces occur and stiction should be avoided.

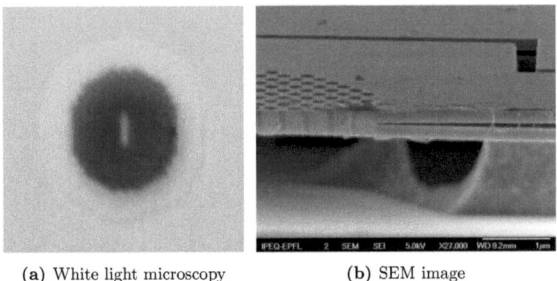

(a) White light microscopy (b) SEM image

Figure 3.41: White light (a) and scanning electron (b) microscopy after critical point drying techniques. We observe that the membranes are sticking at the location of the photonic crystal. In (a), the color of the cavity, which is different from the color of the halo, is a further indication of sticking.

We then performed SEM on the sample. The inter-membrane wet etching did not reach the same extent as for sample P920, although the time was slightly increased (5 and 7 minutes). Unfortunately, we also observe a clear sticking in the region of the photonic crystal on all observed cavities, while other structures seem to be well separated. White light microscopy confirms the result. For example, on figure 3.41a, the center of the cavity has not the same color as the halo, which is an indication of cavity sticking.

We are currently performing further experiments with critical point drying techniques. We are also considering setting the top membrane in a low tensile pre-strain, for example by incorporating In atoms in the first few nanometers of the top membrane. As the inter-membrane $Al_{0.7}Ga_{0.3}As$ layer is etched away and the cantilever structure is released, the top membrane will tend to bend upwards to relax the elastic strain due to the larger material lattice constant of InGaAs [Vaccaro 01].

We are also aiming at an electrical control over the inter-membrane distance (see figure 3.42), that would allow for example to steer the cavity resonance on a time-scale never reached so far with other tuning strategies.

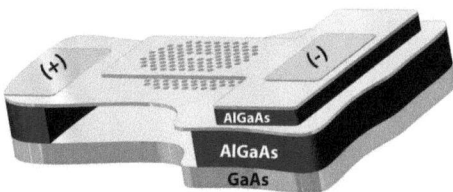

Figure 3.42: Illustration of a device for the electrical control of the inter-membrane distance.

Conclusions

The double membrane structure is a very promising tool to study, among other things, the interaction between a light emitter and modes of a cavity. Due to the impressively large range of tuning predicted, exceeding 100 nm, one could imagine coupling the same single QD with different modes of the cavity and even removing the cavity effect by sweeping the band gap away from the dot energy. This could be used, for example, to probe the Purcell effect or, provided fast electrical tuning is achieved, to increase the resonant absorption (similar to[Nomura 06]) in an excited state of the QD.

So far, we demonstrated controlled 3.5 nm blueshift and 6 nm redshift of the cavity mode without notable degradation of the quality factor Q. These results are really encouraging, when compared to other tuning strategies, even though they seem far away from the promised features of the double membrane structures.

3.9 Electric field tuning

While the previous strategies applied principally to the tuning of the cavity mode, we can also shift the exciton energy. We already saw that, by changing the temperature of the sample, we could control the QD emission. However, the useful range for the Purcell effect is limited by the line broadening provoked by interaction with the phonons.

An electric field F applied perpendicularly to the membrane will also displace the energy levels in the QDs and cause a red shift. This is known as the quantum confined Stark effect (see for example chapter 8.C from [Rosencher 02]).

We designed a new membrane sample in a p-i-n configuration. The intrinsic region

3.9. Electric field tuning

with low density quantum dots in its middle is sandwiched between two AlGaAs barriers to confine the free carriers in the undoped region. The top layer is p-doped with Si atoms and the bottom n-doped with Be. The device fabrication details are provided in section 5.2 on page 115 where the structure is used in forward bias as a LED.

Here, we will focus on the reverse bias operation of the device. The measurements were mainly performed by Dr. Nicolas Chauvin at TU/e.

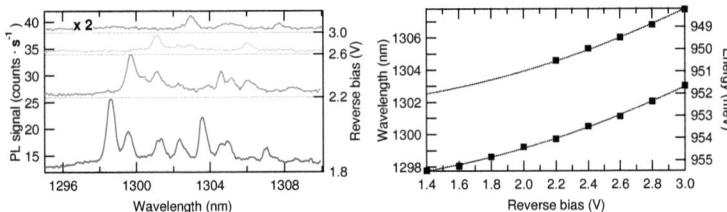

Figure 3.43: (a) QD photoluminescence for 630 nm CW excitation (mW range) under various reverse bias voltages (left axis). The spectra are shifted vertically for clarity. (b) Position of 2 excitonic lines as a function of the applied reverse bias. The energy axis (right) is inverted.

We studied a photonic crystal cavity with a mode at 1290 nm with Q=740. The sample is kept at 4 K in the cryogenic probe station presented in section 2.2.3. We biased the device with increasing reverse voltages and could observe a red shift of the excitonic lines (figure 3.43). As the collection efficiency of the probe-station is limited, we used a low resolution grating to disperse the light in the monochromator and long integration times, between 3 and 5 minutes. Small fluctuations in the applied electrical field and the low resolution detection explain the broad appearance of the excitonic lines on figure 3.43.

Unfortunately, as the cavity mode is on the blue side of the QD lines, it was not possible to control the coupling of the QDs to the mode. Still, the observed tuning range accessible with the quantum confined Stark effect is really interesting, as it spans over approximately 5 nm for a reverse bias range of between 1.4 V and 3.0 V. However, for large voltages, we observe a quenching in the PL.

We repeated the experiment again on another cavity. This time, we used a pulsed laser with wavelength emission at 780 nm and an excitation power in the μW range. The position of two QD lines as a function of the applied field are reported in figure 3.44. The data are

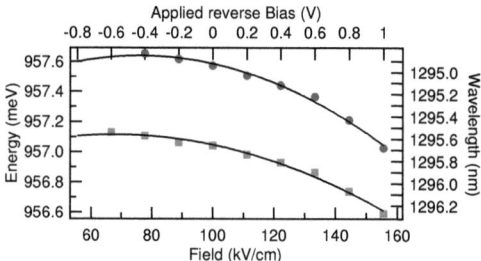

Figure 3.44: Quantum confined Stark shift of 2 QD lines. The solid curves are parabola fitted to the data.

fitted using a parabola [Fry 00]

$$E = E_0 + pF + \beta F^2.$$

The electric field F across the structure is given by $F = (V_{bi} + V_{bias})/d$, where V_{bi} is the built-in voltage of the junction, V_{bias} is the value of the applied reverse bias voltage (we use a minus sign for forward bias) and $d = 180$ nm is the thickness of the intrinsic region. E_0 is the energy of the QD line at $F = 0$, i.e. at the threshold value of the device, which in this case corresponds to $V_{threshold}$=-1.8 V negative bias[k], p corresponds to the exciton dipole moment, and β describes the Stark effect. We obtain $p = (2.3 \pm 0.5) \times 10^{-29}$ C·m for the first QD and $p = (1.4 \pm 0.3) \times 10^{-29}$ C·m for the other QD, which is in good agreement with the $(7 \pm 2) \times 10^{-29}$ C·m reported by Fry et al. for QDs with ground state emission near 1100 nm. As the value of p is positive, it follows that the dipole moment is pointing in the same direction[l] as the field in reverse bias, and we can thus conclude that the electron wavefunction lies below the hole wavefunction for F=0. The amplitude of the Stark effect is given by $\beta = (-9.4 \pm 1.3) \times 10^{-8}$ eV·cm^2·kV^{-2} for the first QD and $\beta = (-6.5 \pm 0.9) \times 10^{-8}$ eV·cm^2·kV^{-2} for the second QD.

In conclusion, by applying an electric field across QDs embedded in a p-i-n device, we were able to successfully red-shift QD lines over more than 5 nm before luminescence quenching. This system also allows for very deep understanding of the QD internal life and further experiments are currently under way.

---

[k] Current starts flowing through the device for a positive bias of 1.8 V.
[l] Per definition, the dipole moment points from the negative to the positive charge.

4

Harvesting of light

Photons emitted by the QDs at the hearth of the GaAs need to be collected for further manipulation. We saw in figure 1.4 that by building a cavity around the emitter, the photons can be extracted from the semiconductor through the mode field pattern. It has been shown for micropillars [Rigneault 01], for example, that the emission of the fundamental mode HE_{11} has a good unidirectionality.

In the first part of this chapter, we will concentrate on improving the vertical extraction properties of the far-field for L3 and H1 cavities, by slightly altering the shape of the photonic crystal defect. We will show that good extraction is possible while keeping large Q values. In a second time, we will explore another approach that consist in collecting the cavity photons through a coupled waveguide. This is of major importance for on-chip photonic circuits, but can also be used to extract the photons through the end facets of the semiconductor.

4.1 Vertical extraction

In the weak coupling regime, the radiative emission rate of the QD is increased by the Purcell effect, and the photons are funneled out to free space through the cavity mode, thus bypassing the total internal reflection at the GaAs/air interface (see figure 1.4). To achieve efficient coupling into a single mode fiber using conventional optics, the farfield of the mode should be both well shaped and matching the numerical aperture of the collecting lens. A Gaussian-like farfield is preferable to have a good overlap with the fundamental mode of the fiber

I will show in the following that we were able to increase the collection efficiency into our optical system, without degrading Q, by tuning the shape of the photonic crystal cavity.

4.1.1 Design

The avoid the tedious optimization of the cavity properties by trial and measurement only, we decided to perform simulations. The models define the parameter space to consider for the sample fabrication. The simulations were performed by Friedhard Römer[a] with a 3D finite element Maxwell solver (FEM) [Römer 07a]. The calculations were carried out for a triangular photonic crystal with a lattice parameter $a = 340$ nm, hole radius $r_0 = 105$ nm ($F = 35\%$) in a 320 nm thick membrane.

Modified L3 cavities

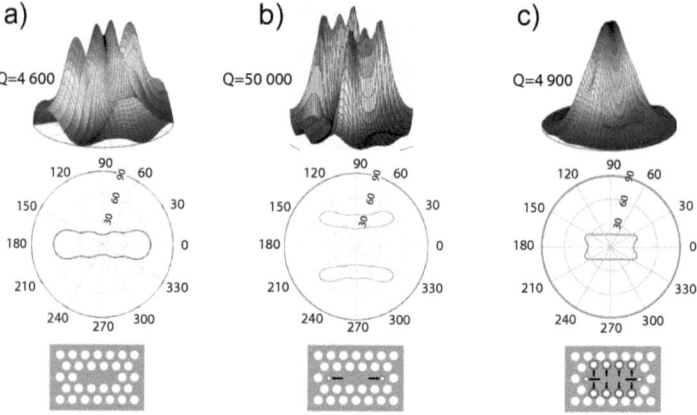

Figure 4.1: Calculated angular farfield intensity of the fundamental mode of (a) an unmodified L3 cavity, (b) a modified L3 cavity optimized for high Q by shifting laterally and scaling the first holes in ΓK direction, and (c) a modified L3 cavity optimized for narrow vertical beaming by outwards shifting the holes in ΓM direction. The polar plot gives the contour at half maximum.

---
[a]Integrated Systems Laboratory, ETH Zurich

4.1. Vertical extraction

Figure 4.1 shows the calculated farfield of the fundamental cavity mode for different L3 cavity configurations. The angular intensity at half maximum is reported on the polar plot below the 3D representation. A microscope objective with a numerical aperture NA=0.5 has a collection angle of 30°, corresponding to the innermost circle on the polar plot.

The farfield of the unmodified L3 cavity is strongly asymmetric, and the calculated Q factor is limited to 4 600 (figure 4.1a). If we compute the modified L3 cavity optimized for high $Q = 50\,000$ by moving the holes outwards along the ΓK direction ($d/a = 0.176$) and reduce their radius to ($r/r_0 = 0.3$), we obtain an emission pattern with two strong lobes emitting between 30° and 60°, far from an ideal Gaussian beam.

We can narrow the emission angle by moving the holes apart in the ΓM direction. The optimized farfield, for a displacement of $d/a = 0.059$, is mainly concentrated within a small divergence angle of 30°. However, its shape is still asymmetric, due to the elongation of the L3 cavity, and on top of that, its Q factor is reduced by an order of magnitude to 4 900.

Modified H1 cavities

The symmetrical shape of H1 cavities motivated us to look closer at their farfield properties. The cavity has a double degenerated dipole mode with orthogonal linear polarization for $F = 35\%$ (see figures 3.2 and B.2). They have a low $Q = 175$ and an emission at half maximum concentrated in a divergence angle of 60°.

Simulations show that the two modes behave in the same way, except for the orthogonal orientation of near- and farfield profiles.

We used the same approach as for L3 cavities and modified the size and position of the first layer of holes. The two dimensional map showing the calculated value for the Q of the mode is drawn on figure 4.2. We observe that for a very narrow range of parameters, Q can be dramatically increased up to 28 500 for $d/a = 0.09$ and $r/r_0 = 0.7$.

The farfield was computed for 5 different configurations and the profile is sensibly different depending on the exact shape of the cavity. For large holes radius (insets a and b), it is mainly concentrated at angles between 30% and 60%. Close to the maximum Q value (inset c), in addition to the wide angle emission, a narrow beam of equal intensity is emerging. For smaller hole sizes (inset d and e), the diverging beams disappear and the emission seems well collimated, with a emission angle below 30°

Clearly, it is not possible to attain an optimal farfield with the highest Q values. However, the simulations indicates satisfying trade-off between a still high value of $Q \simeq 23\,000$

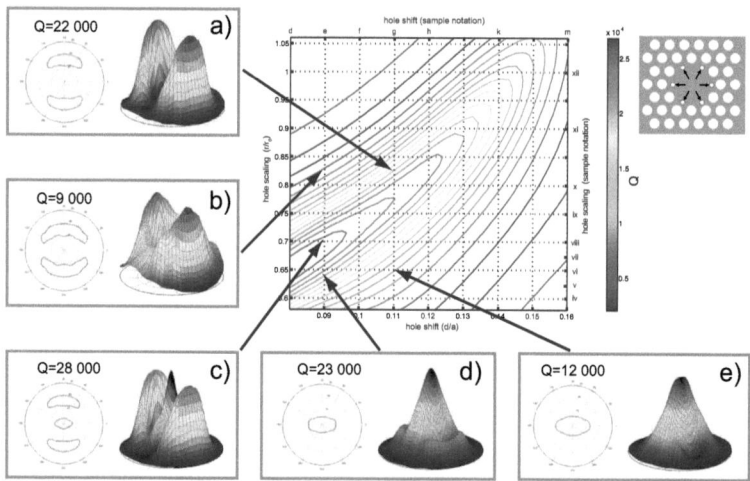

Figure 4.2: H1 cavity modified by scaling and outwards shifting the first layer of holes. The map gives the simulated value of Q as a function of the holes relative shift and radius for the degenerate dipole mode. The insets indicate the angular farfield intensity of one of the polarization for different hole configurations. The polar plot indicates the value at half maximum.

and a reasonable farfield for the efficient collection of light.

4.1.2 Experimental realization

We produced a sample with a high-density of QDs to measure the quality factor Q and the collection efficiency as a function of different first layer of hole shifts:

Name:	a	b	c	d	e	f	g	h	k	m	n
$d/a=$	0.0	0.04	0.06	0.08	0.09	0.1	0.11	0.12	0.14	0.16	0.2

and radius:

Name:	o	i	ii	iii	iv	v	vi	vii	viii	ix	x	xi	xii
$r/r_0=$	0.0	0.4	0.5	0.55	0.6	0.625	0.65	0.675	0.7	0.75	0.8	0.9	1.0

4.1. Vertical extraction

Although high density QDs samples usually show lower Q due to reabsorption, the spectral shape of the mode is clearer to analyze.

Micro-photoluminescence

With our experimental setup, it is not possible to measure directly the farfield of the mode. However, since the microscope objective has a numerical aperture of NA=0.5 and the PL is then coupled to a single mode fiber, it is possible to estimate the collection efficiency by comparing different cavities. The excitation power is kept constant and the cavity is carefully aligned by maximizing the PL signal.

A typical microPL spectrum is shown on figure 4.3a. In the inset, we see that the

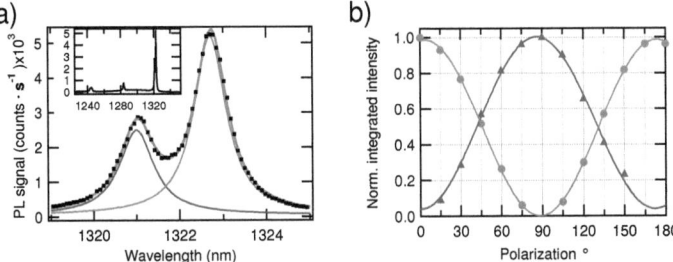

Figure 4.3: (a) PL spectrum of a modified H1 cavity with $d/a = 0.2$ (n) and $r/r_0 = 0.7$ (viii). The dipole mode doublet is well fitted with a sum (green curve) of two Lorentz functions (blue and red curves). (b) Normalized integrated intensity of each mode as a function of the polarizer angle. The background noise was substracted. The modes show intense linear polarization in orthogonal directions.

cavity is becoming multi-mode: as its physical size increases, it can support a quadrupole, a monople, and a hexapole mode. Kim *et al.* extensively studied the hexapole mode and its farfield properties [Kim 06].

We also notice that the degeneracy of the two orthogonally polarized dipole modes is lifted, and two separated peaks are observed. The origin is in structural imperfections of the photonic crystal lattice (see 2.1). To measure the Q factor and the the spectrally integrated intensity of the mode, we carefully fitted each spectrum with the sum of two Lorentzian fit.

$$f(\lambda) = y_0 + \frac{A_1}{(\lambda - \lambda_1)^2 + B_1} + \frac{A_2}{(\lambda - \lambda_2)^2 + B_2} \tag{4.1}$$

The two resonances have orthogonal linear polarizations. Figure 4.3b shows the value of the spectrally integrated intensity as a function of the polarizer orientation. The background noise was removed and the curves are normalized. The continuous lines are sinusoidal fits through the data.

In most of the cavities studied, the doublets have similar width. The average value for the Q factor and the integrated intensity $Area$ are then calculated as:

$$Q = \frac{1}{2}\left(\frac{\lambda_1}{2\sqrt{B_1}} + \frac{\lambda_2}{2\sqrt{B_2}}\right) \qquad Area = \pi \left(\frac{A_1}{\sqrt{B_1}} + \frac{A_2}{\sqrt{B_2}}\right) \tag{4.2}$$

We measured cavities with different lattice parameter a, the position of the mode with respect to the QD emission is thus changed. To correct for this effect, we normalized the data to the maximum intensity for each value of a.

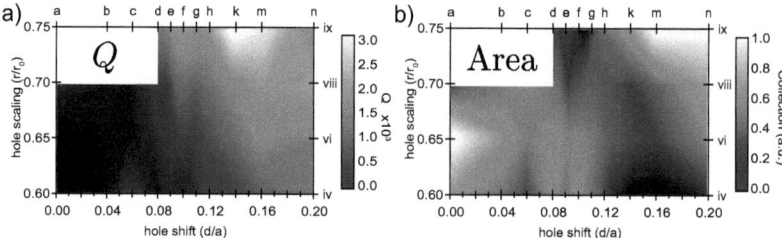

Figure 4.4: (a) Average measured Q and (b) intensity of the dipole mode of a modified H1 cavity for various first holes shifts and scalings.

Figure 4.4a shows the average value of Q as a function of the first layer of hole shift and radius. We notice that the highest value of Q are around 3 000, and that the general shape of the map does not follow the predictions from the simulation.

On figure 4.4b, representing the collection efficiency of our optical setup for different cavity configurations, we also observe discrepancies with the calculations. The region ($r/r_0 = 0.65$ and $0.9 < d/a < 0.11$) where the mode is expected to have the most gaussian-like profile corresponds to the lowest collection.

4.1. Vertical extraction

Still, we notice on the upper right part of the maps, that it is possible to achieve at the same time high collection and high Q.

Fabrication imperfections

To understand the origin of the discrepancy between the simulations and the experimental results, we did an extensive study of the cavity structure with scanning electron microscopy (SEM).

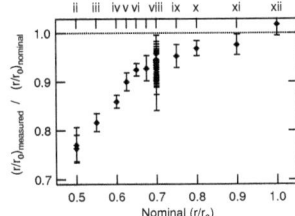 

Figure 4.5: (a) Ratio between the measured and the designed radius for the first layer of holes surrounding the cavity. (b) SEM image of a cavity with nominal parameters $(r/r_0) = 0.625$ and $(d/a) = 0.09$.

The differences between nominal and measured structural parameters for the first layer of holes can be approximated by:

$$\langle d/a \rangle_{\text{meas}} \simeq (-0.007 \pm 0.001) + (1.05 \pm 0.01)\,(d/a)_{\text{nominal}}$$

$$\langle r/r_0 \rangle_{\text{meas}} \simeq (-0.23 \pm 0.03) + (1.25 \pm 0.04) \cdot (r/r_0)_{\text{nominal}}$$

We see that the measured relative shift $\langle d/a \rangle_{\text{meas}}$ is really close to the targeted value $(d/a)_{\text{nominal}}$. However, the measured hole relative dimension $\langle r/r_0 \rangle_{\text{meas}}$ is smaller than expected. This effect can be best seen on figure 4.5a, where we represented the ratio between measured and nominal hole radius. Masks written by ebeam are really precise concerning the relative position of the structures. However, the exposed surfaces are more difficult to control accurately and will depend on the beam intensity and mask thickness. The dry etching steps for the mask transfer may also produce these size deviations from the designed values.

This is even more evident from the measured filling factor. The simulations were performed for $F = 35\%$, but the measured value was $\langle F \rangle_{\text{meas}} = (25.5 \pm 0.1)\%$. This huge difference is responsible not only for a change in the reduced frequency of the mode of $\Delta(a/\lambda) = -0.065$, which was observed on the resonance wavelength, but also in a modification of its field pattern.

With this in mind, it is not surprising that the experimental results did not match the simulations. It also stresses the importance of the fabrication process and the sensitivity of the photonic crystal structures to perturbations. Nevertheless, we could show that high Q factors and good collection efficiency can be compatible by fine-tuning the shape of a H1 defect cavity.

4.2 Cavity in waveguides

Another possible strategy to overcome the loss of collected photons due to total internal reflection and the farfield mismatch to the collecting optics could be to not extract the photons at all from the GaAs, but to design structures where the QDs would emit light principally in the mode of a waveguide and manipulate the photons directly inside the membrane.

4.2.1 Design

If we consider the field pattern of the fundamental mode of a L3 cavity [Akahane 03], we notice that it does not stop abruptly at the edges of the photonic crystal defect, but extends further in the lattice of holes. Would it be possible to modify the shape of a L3 cavity in order to obtain large coupling to a photonic crystal waveguide along the ΓK direction while maintaining a large Q? With this idea in mind, F. Römer performed simulations with the finite element Maxwell solver.

Reducing the number of holes along the ΓK direction increases the coupling to the waveguide at the cost of Q. The effect becomes dramatic between three and two limiting holes. This led to the idea of keeping three holes but alter their size and position to retain a high Q and to maximize the coupling to the waveguide.

Figure 4.6 shows the effect of resizing the middle hole while the two neighboring ones have a radius $r' = \frac{1}{3}r_0$. According to this example, it should be possible to transfer 2/3 of the photons coupled to the mode into the waveguide, while achieving a mode $Q \simeq 20\,000$.

4.2. Cavity in waveguides

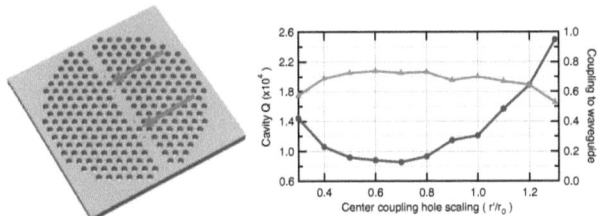

Figure 4.6: Design of a modified L3 cavity optimized for high Q value and good coupling to the photonic crystal waveguide.

4.2.2 Experiment

I realized several designs based on the simulations. We decided to collect the light from the side of the sample through an access waveguide to estimate the lateral cavity coupling to the photonic crystal waveguide. Figure 4.7 is a composite of two SEM images showing the shape of the rib waveguide (left) and its connection to the photonic crystal waveguide (right). We used the layout proposed by Rosa et al. to reduce the reflections at the interface with the photonic crystal waveguide [Rosa 05].

The total length of the photonic crystal waveguide is of 300 periods ($\sim 100\,\mu$m), and we placed the cavity at 10, 50, 100, or 150 periods (approximately 3, 17, 33, or 50 μm) from the access waveguide.

In a first phase, used a sample with high a density of QDs to perform room temperature measurements. The fabrication process is relatively delicate, as it requires careful alignment of the two types of waveguide. The access waveguides are exposed first, and then the membrane is partially wet etched to form the rib. In a second lithographic step, the photonic crystal structure is exposed and we continue the process following the procedure described in section 2.1.

We measured the photoluminescence in the setup presented in section 2.2.4 by exciting the QDs from the top with a 750 nm fiber coupled laser. The luminescence is collected both from the sample surface, as in conventional micro-PL, and from the side through the access waveguide. The light is coupled into single mode fibers and sent to the spectrometer for analysis. The largest cavity quality factors, measured from the top, reach $Q \simeq 3\,600$.

The two spectra presented in figure 4.8 originate from nominally identical cavities, but

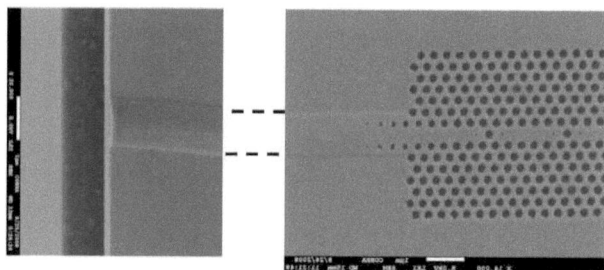

Figure 4.7: SEM image of the device. (left) Cross section image showing the shape of the access waveguide etched in the membrane. (right) Top view of the junction with the photonic crystal waveguide. We see a modified L3 cavity aligned at 10 periods (∼3 μm) from the junction.

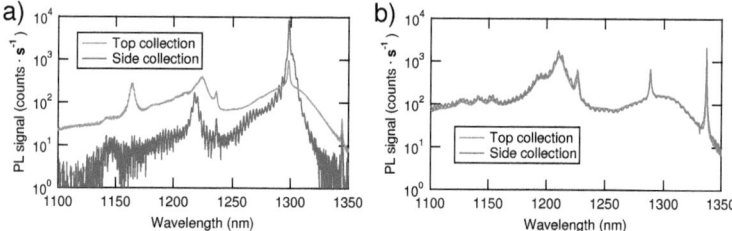

Figure 4.8: Micro-photoluminescence spectra excited from the top of the cavity. The red curves are collected from the top and the blue curves from the side through the access waveguide. (a) Cavity at 10 lattice parameters (∼3 μm) from the access waveguide (see figure 4.7). (b) Cavity at 150 lattice parameters (∼50 μm) from the access waveguide.

positioned at different depth into the photonic crystal waveguide. The signal collected from the side generally shows additional Fabry-Perot resonances due to the alignment mismatch between the photonic crystal and the access waveguides, and from the reflectivity at the cleaved facet. The spectra presented here only show this second type of resonances. We interpret it as a good matching between the two waveguides: when they are aligned correctly, the taper region proposed by Rosa *et al.* effectively reduces the reflectivity at the interface between the waveguides.

4.2. Cavity in waveguides

On figure 4.8a, we see that at the 1300nm resonance, the amplitude of the signal collected from the side is one order of magnitude larger than the one recorded from the top of the sample and that it is otherwise less intense. On figure 4.8b, for the cavity sitting at 150 lattice parameters ($\sim 50\,\mu$m) form the access waveguide, we see that both signals overlap perfectly, except for the small ripple due to Fabry-Perot interference.

As we had difficulties with the fabrication of the sample, we do not have enough measurements yet to interpret these results quantitatively. However, according to figure 4.8a, we see that it is possible to collect more light from the mode coupled through the waveguide than from the surface. This result is really encouraging and we are currently working on a new sample design that is more tolerant to misalignment during the writing of the mask.

5

Control of the spontaneous emission rate

5.1 Single Quantum Dots in a Photonic Crystal nanocavity

In this section, we demonstrate coupling at $1.3\,\mu$m between single excitonic lines from a QD and the optimized fundamental mode of a modified L3 photonic crystal defect cavity with a quality factor $Q \simeq 15\,000$. By spectrally tuning the cavity mode, we observe selective coupling of two closely spaced QDs. We perform time-resolved photoluminescence on one of the excitonic lines and measure an 8-fold increase in the spontaneous emission rate and a coupling efficiency of 96%. This work was published in [Balet 07].

5.1.1 Experimental method

The sample, grown by molecular beam epitaxy, consists of a 320 nm-thick GaAs membrane on top of a $1.5\,\mu$m $Al_{0.7}Ga_{0.3}As$ sacrificial layer. A single layer of low density ($5{\sim}7$ QDs/μm^2) self-assembled InAs QDs emitting at $1.3\,\mu$m at low temperature [Alloing 05] is embedded in the middle of the membrane to maximize the coupling to the fundamental waveguided mode. The fabrication process of the photonic crystal nanocavity is presented in section 2.1.

As a result, we obtain a photonic cristal structure with 7 shells of holes around the nanocavity, at the center of a 12 μm-diameter suspended membrane. The L3 defect cavity presented hereafter (see inset of figure 5.1) is formed by keeping three holes un-etched along the ΓK direction of the triangular photonic crystal. The lattice constant $a = 340$ nm and

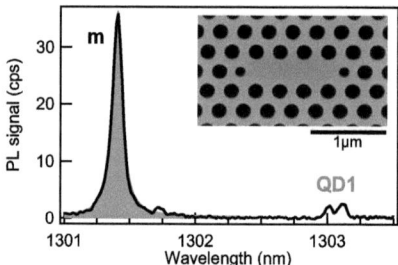

Figure 5.1: MicroPL spectrum of the modified L3 cavity around its fundamental mode (**m**) under pulsed excitation (80 MHz, 80 nW, 750 nm). A Lorentzian fit (grey area) of the mode yields a quality factor $Q = 14\,850$. A doublet excitonic line labeled **QD1** is visible at lower energies. The inset shows a scanning electron microscopy (SEM) image of the cavity.

circular holes with diameter $d = 125$ nm provide a photonic band gap for TE polarization around 1.3 µm. This corresponds to the ground state emission wavelength of our QDs at 10 K. To obtain a smoother mode field pattern, and thus a higher Q, we modified the first holes on both ends of the cavity [Akahane 03]. They are shifted by $d = 0.15\,a$ outwards, and their radius is $r = 2/3\,r_0$, following the experimental optimization presented in section 3.1.

The sample was kept in a liquid Helium continuous flow cryostat and all measurements were performed around 10 K. We used a pulsed laser (pulse width ∼50 ps) at 750 nm for excitation in the GaAs. The luminescence was coupled to the single mode fiber to ensure higher spatial resolution. The experimental setup is described in section 2.2.1. Figure 5.1 shows a typical micro-PL spectrum obtained at an average excitation power of 80 nW. Interestingly, the cavity mode is visible, even when it is spectrally far from excitonic lines. We attribute this emission to an energetically broad multiexcitonic background emission correlated to the QD [Hennessy 07], most probably originating from the interaction with randomly fluctuating environmental charges [Kamada 08]. The cavity mode has a Lorentzian lineshape. Its full-width at half maximum of 64 µeV yields a quality factor $Q = E_0/\Delta E = \omega_c/\Delta \omega_c \simeq 14\,850$, which is an indication for the high quality of the sample processing, as it is the state of the art for cavities containing QDs. This is also promising for the control of spontaneous emission since, following equations (1.11) and (1.14), we can

5.1. Single Quantum Dots in a Photonic Crystal Nanocavity

expect a maximum achievable rate enhancement:

$$\frac{\tau}{\tau_0} = \frac{1}{2} \cdot F_p = \frac{3Q(\lambda_c/n)^3}{8\pi^2 \mathcal{V}_{cav}} \simeq 780, \quad (5.1)$$

assuming [Akahane 03] a mode volume $\mathcal{V}_{cav} = 0.73(\lambda_c/n)^3$, or even see strong coupling. However, as the position of the emitter is random and in general not perfectly matched to the cavity field, we expect to operate in the weak-coupling regime with a lower Purcell factor.

A doublet emission line, labeled QD1 is visible at 1303.0 nm. Based on temperature- and power-dependent measurements, these line can be definitely ruled out as cavity mode. Micro-PL as well as time-resolved measurements indicate that both peaks originate from the same spatial location and we attribute this doublet structure to the splitting of the bright exciton angular momentum states due to QD asymmetry [Bayer 02b]. The same observation holds for QD2 at 1303.8 nm.

5.1.2 Detuning

Temperature tuning is commonly used to spectrally bring the QDs in resonance with the cavity since they red shift faster than the mode. However, the temperature increase changes the relative intensity of the various excitonic lines of a single QD and causes line broadening (figure 1.2), thus lowering the rate enhancement, as we saw on equation (1.14). This also makes the analysis of the micro-PL of the QD to cavity coupling more difficult. Moreover, this particular QD lines are already on the red side of the mode. We used the gas deposition method presented in section 3.3. Spectra taken at different times clearly show the red-shift of the mode and its crossing over the QDs excitonic lines (figure 5.2a). As the energy detuning between the mode and the QDs decreases, we observe an enhancement of the integrated micro-PL signal from the excitonic lines. This clearly indicates coupling with the cavity mode: photons emitted by the exciton efficiently escape the GaAs through the mode emission pattern. However, as we operate under pulsed excitation, this is not an indication of increased spontaneous emission rate. No splitting is observed at resonance, indicating that despite the high Q value, the cavity–QD system is not in the strong coupling regime, probably due to non-optimal spatial coupling [Badolato 05]. The integrated intensity of the exciton as a function of the detuning (figure 5.2b and 5.2c) was obtained by fitting the region of interest in the micro-PL spectra with the sum of a Lorentzian for the mode, and two Gaussian for each QDs doublet. It follows a Lorentzian trend, as expected in the case

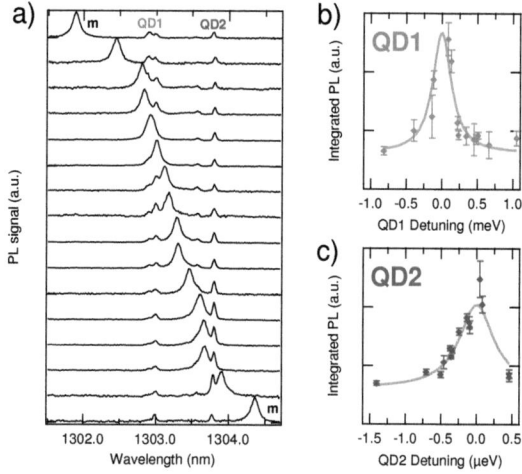

Figure 5.2: (a) μPL spectra (15 μW, 80 MHz) showing two excitonic lines, QD1 at 1303.0 nm and QD2 1303.8 nm, and the red shift of the mode (m) due to air molecules being cold-trapped on the surface of the sample (see section 3.3). As the mode approaches each QD line, we observe a marked increase in its emission intensity. (b) Integrated intensity of QD1 as a function of the detuning with respect to the mode energy. The width of the detuning curve is 316 μeV. (c) Integrated intensity of QD2 as a function of the detuning with respect to the mode energy. The width of the detuning curve is 466 μeV.

of Purcell enhancement.

5.1.3 Time-resolved measurements

To confirm that this emission enhancement is due to the Purcell effect and to quantify it, we performed time-resolved measurements on QD1. The micro-PL signal is spectrally filtered by a tunable filter with 0.8 nm bandwidth and directed to a fiber-coupled SSPD (see section 2.2.1). The histogram of delays between the laser pulse and the detector output is recorded. The SSPD has a quantum efficiency of 10 % and a dark count rate of 8 Hz at a bias current of 21 μA. The temporal resolution of the system is around 150 ps and deconvolution was performed to extract the emitter lifetime. The sample was excited with ∼5 nW average

5.1. Single Quantum Dots in a Photonic Crystal nanocavity 113

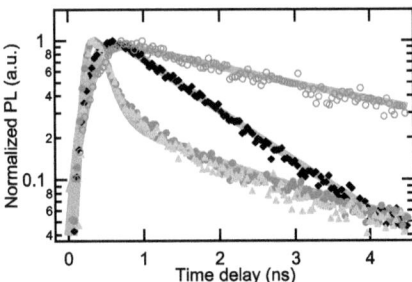

Figure 5.3: Time resolved dynamics of QD1 off resonance (empty circles), of ensemble QDs outside the photonic crystal patterned region (diamonds), and of QD1 on resonance with the cavity mode under 5 nW (filled circles) and 0.4 nW (filled triangles) excitation power and 20 MHz repetition rate.

power at 20 MHz. The measured lifetime of QDs away from the cavity (figure 5.3, diamonds) is $\tau^{QD} \simeq 1.2$ ns. This is in good agreement with previously reported values for the exciton lifetime in single QDs grown under the same conditions [Zinoni 06]. For QD1, in the cavity, but energetically off-resonance, we observe a suppression of the spontaneous emission rate, as shown in figure 5.3 (empty circles), where, for a 2.5 nm detuning, the lifetime of QD1 is $\tau_{\text{off}}^{QD1} \simeq 3.6$ ns. In fact, when the cavity mode is off the QD line, the density of states of electromagnetic modes at the QD energy is smaller than in free space resulting in an inhibition of the spontaneous emission and thus a slower decay rate through the leaky modes, as discussed in section 1.2.2 and observed already [Gayral 01, Gevaux 06, Chang 06]. The PL decay time of the QD on resonance (figure 5.3, filled circles) clearly shows two time constants. The fast one, $\tau_1^{QD1} \simeq 150$ ps, is attributed to the Purcell effect and indicates a $(\tau^{QD}/\tau_1^{QD1}) \simeq$ 8-fold rate enhancement. The slow time constant, $\tau_2^{QD1} \simeq 1.9$ ns, can be attributed to the spin flip transition from dark to bright exciton [Chang 06], or to the spectrally broad background emission coupled into the filter bandwidth.

However, to exclude this last possibility, we performed the experiment at several pumping powers. Both the exciton and the background couple to the cavity line in the resonance condition, but since the background emission presents a stronger power dependence than the sharp excitonic line it is possible to discriminate the two effects and ensures that most of the signal is indeed originating from the QD1 line. Since the same curve is obtained for different

114   Chapter 5. Control of the spontaneous emission rate

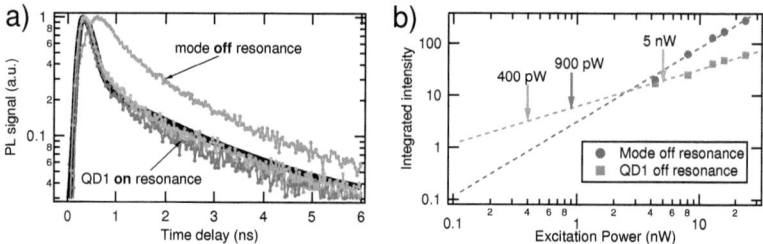

Figure 5.4: (a) Intensity normalized time-resolved photoluminescence of QD1 line in resonance with the mode for 400 pW, 900 pW and 5 nW excitation. The dark solid line is a convolution of a bi-exponential curve with the temporal response of the setup fitting the 5 nW excitation curve. The slower dynamics is the mode off resonance for a 1.3 nW excitation power. (b) Comparison between the intensity of the excitonic line off resonance and the mode off resonance at different powers based on a simple linear model.

low powers (figure 5.4a), we can conclude that the background is efficiently suppressed with respect to the exciton luminescence. In contrast, the mode measured off resonance with the exciton line, but pumped by the background emission, presents a different dynamics (grey line in figure 5.4a) with a deconvoluted lifetime of 370 ps.

The coupling efficiency into the mode, defined in section 1.2.3 as $\beta \simeq 1 - (\tau_{\text{cav}}/\tau_{\text{leak}}) = 1 - (\tau_1^{\text{QD1}}/\tau_{\text{off}}^{\text{QD1}})$ is $\sim 96\,\%$. This result is very promising: as most photons are emitted into the cavity mode, a very large collection efficiency can be obtained by optimizing the farfield of this mode (chapter 4).

5.1.4 Detuning and effective Purcell factor

In a recent publication [Gayral 08], the authors studied hidden pitfalls in photoluminescence experiments performed on high Purcell factor microcavities. Although they assume a dense distribution of QDs, which does not hold in our case, we can still use some of their conclusions in this chapter. They demonstrate that the collected signal from the mode has a Lorentzian lineshape with a width

$$\Delta_{QD} = \frac{E_0}{Q}\sqrt{\frac{E_p + \gamma}{\gamma}}, \qquad (5.2)$$

which appears broadened by the contribution of multiple QDs excitons. In our case, this corresponds to the detuning curves presented in figure 5.2b) and c). $E_0 = 0.951\,\mu$eV is the energy of the mode on resonance, $E_p = \frac{\tau_0}{\tau_{cav}}$ is the effective Purcell enhancement, and $\gamma = \tau_0/\tau_{\text{leak}} = 0.33$.

If we evaluate equation 5.2 for QD1 with $E_p=8$, we expect a width of 322 μeV, in excellent agreement with the measured $\Delta_{QD1} = 316\,\mu$eV. By solving the equation again for QD2 with a measured width $\Delta_{QD2} = 466\,\mu$eV, we obtain an effective Purcell enhancement of 17. This results in an expected on-resonance lifetime of barely 70 ps, well below the temporal resolution of our system.

5.1.5 Conclusion

In summary, we presented modified L3 defect nanocavities on GaAs with very high Q of 15 000 at 1 300 nm wavelength. We observed a clear enhancement of micro-PL signal of single excitons as a function of the energy detuning to the cavity mode. We further demonstrated Purcell effect on this cavity by time-resolved analysis of the dynamics of two neighboring transition at 1.3 μm. We directly measured an 8-fold enhancement of spontaneous emission rate yielding a $\beta \simeq 96\%$ coupling efficiency. A second transition even shows a larger Purcell effect, which is estimated to be around 17.

The reduced QD lifetime may lead to coherent, transform-limited single photon sources. Together with the increased extraction efficiency, it pledges the realization of efficient photonic crystal nanocavities based single-photon emitters at 1.3 μm for optical quantum information processing. The next natural step is the integration in electrically operated devices. This is the subject of the following section.

5.2 Photonic crystal LED

In this section, we present the direct measurement of enhanced spontaneous emission in a photonic crystal membrane nanocavity under electrical operation. The main results are published in [Francardi 08]. The device consists of p–i–n heterojunction embedded in a suspended membrane, comprising a layer of self-assembled QDs. A modified L3 photonic crystal defect nanocavity with fundamental emission around 1.3 μm is etched through the membrane.

Recently, Hofbauer *et al.* measured a 3.5× enhancement of the radiative lifetime by photocurrent spectroscopy on photonic crystal membrane LED [Hofbauer 07]. However, their layout does not allow to address a single cavity and the large dimension probably limits its operating speed. Our original design allows to address single photonic crystal cavities electrically with low parasitic resistance and capacitance. It allows to perform time-resolved electroluminescence measurements, with switch off times shorter than the bulk radiative and nonradiative lifetimes.

5.2.1 Device fabrication

We first considered the design proposed by Park et al. where the injection of holes is accomplished by a pillar just underneath the center of the nanocavity [Park 04]. However, the technological challenge and the concern of degrading the Q of the cavity encouraged us to search for another approach.

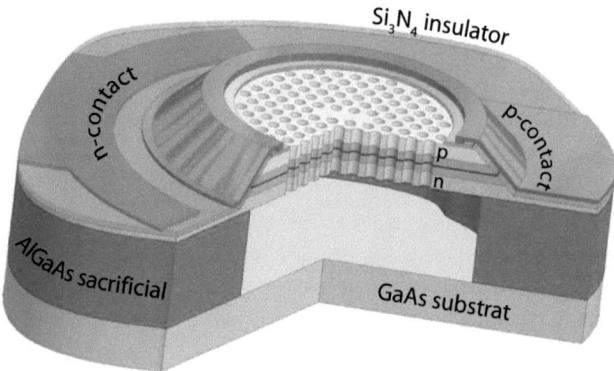

Figure 5.5: Sketch of the LED showing the p-i-n heterojunction suspended membrane.

Our device, schematically shown on figure 5.5, consists of a 370 nm-thick membrane photonic crystal cavity containing a p-i-n heterojunction. The intrinsic region with low density quantum dots ($\sim 7\,\mu m^{-2}$) in its middle is sandwiched between two AlGaAs barriers to confine the free carriers in the undoped region. The top layer is p-doped with Si atoms and the bottom n-doped with Be. The holes are injected in the p-doped layer with a ring-shaped top contact, and the electrons directly in the bottom layer of the membrane through the side of the mesa.

5.2. Photonic crystal LED

Table 5.1: Structure of the sample P945 used to electrically control the QD luminescence.

Material	Thickness	Doping	Comments
air			
GaAs	30 nm	p, $2 \cdot 10^{19}$	top p contact
GaAs	50 nm	p, $2 \cdot 10^{18}$	
$Al_{0.1}Ga_{0.9}As$	10 nm	p, $2 \cdot 10^{18}$	confinement barrier
$Al_{0.2}Ga_{0.8}As$	10 nm	p, $2 \cdot 10^{18}$	confinement barrier
$Al_{0.2}Ga_{0.8}As$	20 nm		confinement barrier
GaAs	65 nm		
InAs QDs	~10 nm		low density, 1300 nm at 4 K
GaAs	65 nm		
$Al_{0.2}Ga_{0.8}As$	20 nm		confinement barrier
$Al_{0.2}Ga_{0.8}As$	10 nm	n, $2 \cdot 10^{18}$	confinement barrier
$Al_{0.1}Ga_{0.9}As$	10 nm	n, $2 \cdot 10^{18}$	confinement barrier
GaAs	40 nm	n, $2 \cdot 10^{18}$	
GaAs	40 nm	n, $5 \cdot 10^{18}$	bottom n contact
$Al_{0.7}Ga_{0.3}As$	1500 nm	n, $2 \cdot 10^{18}$	sacrificial layer
GaAs	1000 nm	n, $2 \cdot 10^{18}$	
GaAs substrate			

The fabrication process is based on several e-beam lithography steps and typical thin-film processes. We first define 8 μm-diameter, 320 nm-deep circular mesas by wet etching down to the bottom n-contact layer. Then, a bow-shaped n-contact is defined by lift-off of a 155 nm-thick Ni/Ge/Au/Ni/Au multilayer that is subsequently annealed at 400°C for 30 minutes. A 200 nm-thick Si_3N_4 layer is deposited to isolate the n and p contacts and then removed from the n-contact surface and from the top of the mesa. By lift-off of a 110 nm-thick Cr/Au bilayer we realize an annular p-contact on the top of the mesa. In the same evaporation step, the top p-contact and the bottom n-contact are connected to ground-signal-ground coplanar electrodes on top of the Si_3N_4 surface. An annular gold cover is evaporated around the entire mesa edge in order to filter out light scattered by the mesa sidewalls. A 150 nm-thick SiO_2 layer is then deposited and the photonic crystal pattern with optimized L3 cavities, aligned to the mesa, is defined by electron beam lithography following the usual procedure (see section 2.1).

Figure 5.6a shows an optical microscopy image of a single LED. The size and shape of the golden pads match the shape of the electrical probe. The blue layer is the insulating Si_3N_4. The scanning-electron microscope (SEM) image of the LED at the end of the fabrication process gives an idea of the dimensions and on the small alignment tolerances allowed for the device.

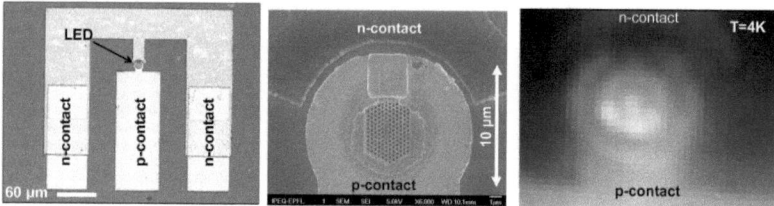

Figure 5.6: (a) Optical microscopy image of a single photonic crystal cavity LED. (b) Scanning electron microscopic image of a 8 μm-diameter device. The p-contact is clearly visible on the bottom side, whereas the n contact is shaded by the Si_3N_4 insulating layer. (c) Infrared ($\lambda > 1\,250\,nm$) image of the device under electrical operation at liquid helium temperature.

5.2.2 Electroluminescence under continuous bias operation

The device was tested in the cryogenic probe station presented in section 2.2.3. Figure 5.6c shows an image of the LED under electrical operation at liquid helium temperature. The image was recorded with an InGaAs camera equipped with a $\lambda > 1\,250\,nm$ long pass filter. We clearly recognize luminescence from the cavity, at the center of the device, but also from the border, near the metallic contacts. An electroluminescence spectrum as well as the LED current-voltage (I–V) characteristic are shown in figure 5.7a and 5.7b.

A clear cavity mode is visibe around 1329nm, superposed over a broad background signal corresponding to ground-state emission from other QDs emitting within the collection area. The full-width half-maximum of the mode peak is 0.29nm corresponding to a quality factor $Q \simeq 4\,600$, similar to the $Q \simeq 4\,000$ measured by photoluminescence in open-circuit.

On figure 5.7c, we plotted the mode intensity as a function of the bias current (L–I curve). The luminescence of the mode was filtered through a 0.2nm tunable bandpass filter

5.2. Photonic crystal LED

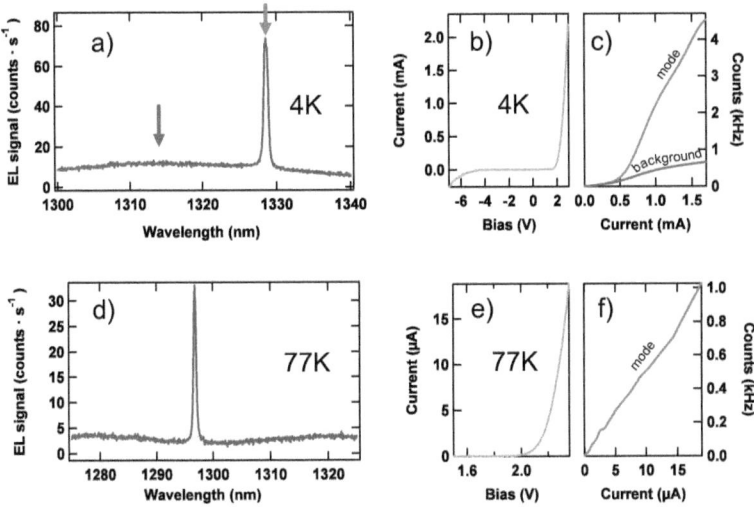

Figure 5.7: *Liquid Helium operation*: (a) Electroluminescence spectrum at a bias current of 1.25 mA, (b) I–V characteristic of the device exhibiting diode operation, and (c) electroluminescence as a function of the injected current (L–I) at the wavelength of the mode (red curve) and detuned from the mode (grey curve) corresponding to the arrows in a.
Liquid Nitrogen operation: (d) Electroluminescence spectrum under 15 μA bias, (e) I–V characteristic of the device exhibiting diode operation, and (f) electroluminescence as a function of the injected current (L–I) at the wavelength of the mode recorded on another LED.

and recorded on the SSPD (section 2.2.1). We see a clear threshold behavior in the light intensity at 0.5 mA, but we do not attribute it to lasing. Indeed, if we compare the L–I curve when the filter is detuned by 15 nm from the mode (at the position of the grey arrow in figure 5.7a), we notice that they follow the same trend up to 0.5 mA. Moreover, the infra-red imaging of the LED (figure 5.6c) shows that at low current bias the emission arises from the region close to the contacts. It emerges from the center of the LED, i.e. from the cavity, only after the bias is further increased. We attribute it to carrier freeze-out: the positive charge carriers (the holes) do not spread fast across the membrane and tend to recombine with the electrons near the p-contact electrode. When the current gets larger, the increased number of charges beats the recombination rate and the holes finally reach the cavity in the center of the membrane.

We repeated the experiment at liquid nitrogen temperature, since this correspond to a maximum in carrier mobility [Hillmer 89]. The electroluminescence spectrum recorded at 77 K and displayed in figure 5.7a is from another device with similar characteristics. If we look at the L–I curve of the mode (figure 5.7f), we notice that the threshold-like feature due to the background disappeared. We also notice that the same electroluminescence intensity on the mode, is reached with 20 times less bias current at 77 K than at 4 K. The higher mobility of the holes allow them to beat the recombination rate and reach the center of the cavity already for low bias currents.

5.2.3 Time-resolved electroluminescence

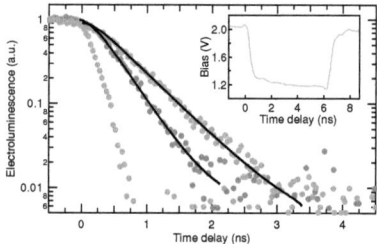

Figure 5.8: Time-resolved electroluminescence of the wetting layer (green), of the mode (red) and detuned from the mode (blue), under 40 MHz pulsed excitation. The electrical pulsed waveform is shown in the inset. The measured average current is 1 mA.

We performed time-resolved electroluminescence at 4 K on a device with a cavity mode at 1319 nm and a quality factor $Q \simeq 1\,000$. We biased the LED with pulsed excitation by combining a constant voltage with a 40 MHz square signal through a bias-T (see figure 2.8). The measured signal at the output port of the bias-T is displayed as an inset in figure 5.8. The diode is biased above threshold during 3/4 of the cycle. The voltage then drops rapidly well below threshold, switching the LED off and allowing the QDs to relax through spontaneous emission. The measured average current flowing through the diode is 1 mA. The electroluminescence signal is sent to a SSPD for detection, and the time of arrival of the photons is recorded by a correlation card synchronized to the pulsed bias.

5.2. Photonic crystal LED

The temporal resolution of the setup was measured form the emission of the wetting layer (figure 5.8, green curve) whose lifetime is expected to be well below 100 ps. The measured decay time of ~ 210 ps is mainly determined by the jitter of the SSPD and of the correlation card. This confirms that device parasitics are low and do not limit the dynamic behavior of the LED.

The temporal decays of the cavity mode (red dots) and QD ensemble energetically detuned off the mode (blue dots) are shown in figure 5.8. The time constants are calculated by fitting the experimental results through the convolution of a mono-exponential curve with the response function of the setup, i.e. the wetting layer dynamics (green dots). Recombination lifetimes of $\tau_{on} \simeq 380$ ps and $\tau_{off} \simeq 580$ ps are obtained for the on-resonance and off-resonance emissions respectively. Thus, a 50 % increase of the recombination rate is observed on resonance with the mode. We note that the off-resonance emission is faster than the typical exciton lifetime of similar QDs in bulk, which is around 1 ns [Zinoni 06]. This could be due to non-radiative effects and/or to emission from multi-excitons. The current flowing through the diode also induces a heating of the device.

5.2.4 Conclusion

In conclusion, we have directly measured the enhancement of spontaneous emission dynamics in a photonic-crystal quantum dot LED. Our design allows for a modulation speed and turn-on delay not limited by the free space spontaneous emission time. Under reverse bias (see section 3.9), this device also allows for fast tuning of the exciton energy through the quantum confined Stark effect.

6

Conclusion

Thanks to the recent development and optimization of a novel growth technique, single quantum dots were available with an emission wavelength in the 1.3 μm range. Antibunching experiments showed the highly non-classical nature of the light emission, making these quantum dots perfect candidates for single photon sources at telecommunication wavelengths.

However, the radiative lifetime around 1 ns combined with the low photon extraction efficiency, due to the high refractive index contrast between the growth material and air, spoils the aforementioned qualities.

The aim of this thesis work was to increase the emission efficiency of single quantum dots by incorporating them in photonic crystal micro-cavities. Coupled emitters undergo faster radiative recombination rates, known as the Purcell effect, and the photons are emitted in the mode farfield, thus bypassing the total internal reflection barrier.

First, we developed the photonic crystal on membrane processing and reached quality factors $Q > 16\,000$ for defect cavities in GaAs. We then explored different strategies to tune the mode and the quantum dots in resonance. Global temperature tuning shifts both the mode and the QDs, but induces an homogenous broadening due to phonon interaction. Local heating, with a focussed laser or a SNOM can modify the refractive index of the cavity without perturbing the QDs or degrading the Q significantly. We also investigated modifications of the effective index by deposition of frozen gas or polymer on the surface of the sample. By approaching a SNOM tip, we were able to induce a local modification of the effective index, shifting the energy of the mode and allowing for high resolution imaging of its electric field pattern. We then obtained first experimental results by developing a technique with a double membrane photonic crystal cavity configuration. The resonance frequency of the mode can be tuned over a wide range and is controlled by the inter-membrane distance. The next step would be to control this distance electrically. We used a p–i–n junction in reverse bias to control the energy of the the exciton through the quantum confined Stark effect. This method has the advantage of being fast and acting on the QD wavelength without inducing a noticeable line broadening.

We then addressed the collection efficiency of the emitted photons by a single mode fiber, both numerically and experimentally, and showed that this is not incompatible with high Q factors.

We observed the Purcell effect on a single QD coupled to the cavity mode by directly measuring an 8-fold enhancement in the spontaneous emission rate and a 96% coupling efficiency under optical pumping at low temperature. As practical devices need to be driven electrically, we then investigated an original design to create a photonic crystal membrane LED. After several optimization steps, we could see cavity modes and measured a 50% increase in the emission rate at resonance under pulsed forward bias.

As we see, they are several approaches to control the frequency coupling between an emitter and a cavity mode. However, to achieve reproducible devices, we need to find strategies to ensure a controlled spatial coupling as well, as this is the key to enter the strong-coupling regime. Among the proposed approaches, the most promising for large scale application would be the control of the QDs nucleation site during the growth procedure.

A

Material dispersion in the master equation

When we calculate the band structure of a photonic crystal cavity with the plain wave expansion (PWE) method, we obtain the results in the form of the reduced frequency $\tilde{\nu}$. The unit of $\tilde{\nu}$ is $(\omega a/2\pi c)$ [Joannopoulos 95], or simply a/λ, with λ the vacuum wavelength. This is a direct consequence of the scalability of the master equation

$$\nabla \times \left(\frac{1}{\varepsilon'(\mathbf{r})} \nabla \times \mathbf{H}(\mathbf{r}) \right) = \left(\frac{\omega}{c} \right)^2 \mathbf{H}(\mathbf{r})$$

which contains no fundamental length scale. One might then fix the lattice parameter a to obtain the wavelength of a mode $\lambda = a/\tilde{\nu}$. Although this approach gives a good approximation, it overlooks the material dispersion. For example, we can compute the modes of a H1 cavity in a 320 nm thick membrane for $\lambda_0 = 1300$ nm. The effective refractive index is $n_{\text{eff}}^{1300} = 3.0949$ and we obtain a single mode at the reduced frequency $\tilde{\nu}^{1300} = 0.27061 \, (a/\lambda)$. Now, if we assume that we have a lattice parameter $a = 319$ nm, we calculate straightforwardly a resonant wavelength $\lambda_r^{1300} = 1179$ nm. Let us perform this calculation again, starting at a wavelength $\lambda_0 = 1200$ nm. In this case, $n_{\text{eff}}^{1200} = 3.1533$, $\tilde{\nu}^{1200} = 0.266 \, (a/\lambda)$, and for $a = 319$ nm, we get $\lambda_r^{1200} = 1199$ nm: a 20 nm discrepancy! Table A.1 illustrate this on another example.

The effect if also visible while calculating the resonances frequencies of a multi-mode cavity. The calculated position of the peaks will not match the measured one if we overlook the material dispersion while solving the master equation.

125

λ_0 (nm)	$n_{\text{eff}}(\lambda_0)$	$\tilde{\nu}(a/\lambda)$	a (nm) for $\lambda_r = 1300$ nm
1500	2.9900	0.27929	363
1400	3.0407	0.27503	357
1300	**3.0949**	**0.27061**	**352**
1200	3.1533	0.26599	346
1100	3.2180	0.26105	339

Table A.1: This table illustrates the effect of neglecting the material dispersion while estimating the resonance wavelength of a mode with the PWE method. The task is to calculate the lattice parameter to have the mode of a H1 cavity at 1300 nm. The effective refractive index $n_{\text{eff}}(\lambda_0)$ is calculated for the fundamental mode of a 320 nm thick GaAs membrane at the wavelength λ_0 while taking into account the material dispersion. The third column is the reduced frequency of the mode of a H1 cavity calculated by the PWE method with the effective index from the second column. The fourth column is simply the photonic crystal lattice parameter $a = \lambda_r \cdot \tilde{\nu}$ with $\lambda_r = 1\,300$ nm.

B

Modes of selected photonic crystal cavities

This appendix presents the structure of the modes for various photonic crystal nanocavities. It helps to understand the polarization properties of the emitted light. The simulations were performed using the plain wave expansion (PWE) routine developed by Vasily Zabelin in the frame of his PhD thesis in the group of Romulad Houdré at EPFL. The filling factor of the photonic crystal is 35%, corresponding to a value $r/a \simeq 0.31$, and the refractive index was chosen at $n_{\text{eff}} = 4.00$. In figures B.1, B.3, and B.5, the dashed lines represent the top and bottom band of the photonic crystal bandgap, the squares are simulation results obtained with the PWE code, using the primitive cell shown on figure B.2, and the solid lines are a polynomial fit trough the data.

Chapter B. Modes of selected photonic crystal cavities

B.1 H1 cavity

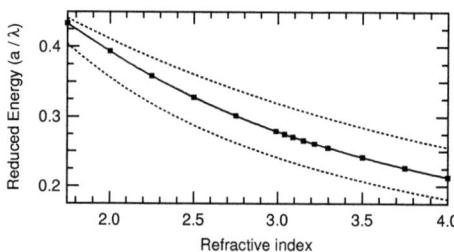

Figure B.1: Reduced energy of the double degenerated mode of the H1 cavity as a function of the refractive index.

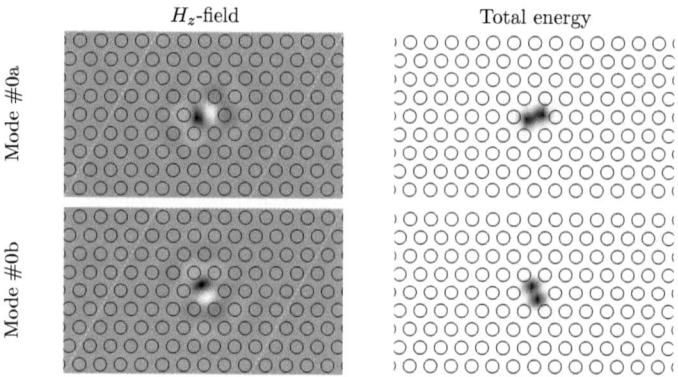

Figure B.2: The degenerate dipole mode of the H1 cavity for $n_{\text{eff}} = 4.00$.

B.2 L3 cavity

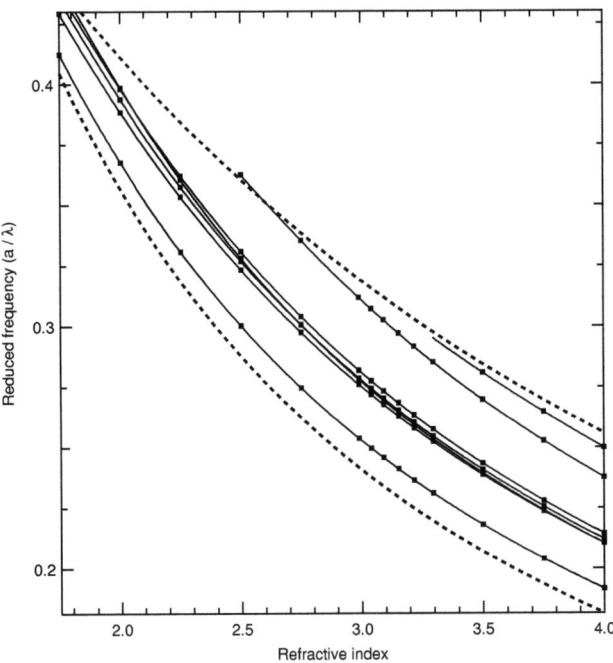

Figure B.3: Reduced energy of the modes of the L3 cavity as a function of the refractive index.

130   Chapter B. Modes of selected photonic crystal cavities

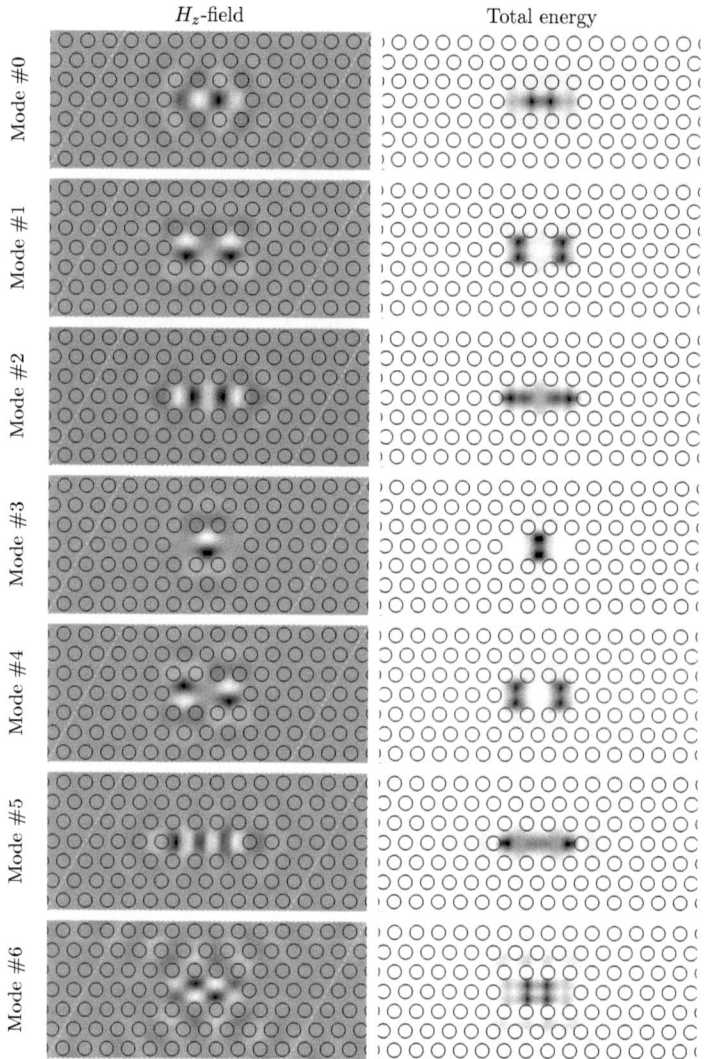

Figure B.4: The modes of the L3 cavity for $n_{\text{eff}} = 4.00$.

B.3 Modified L3 cavity

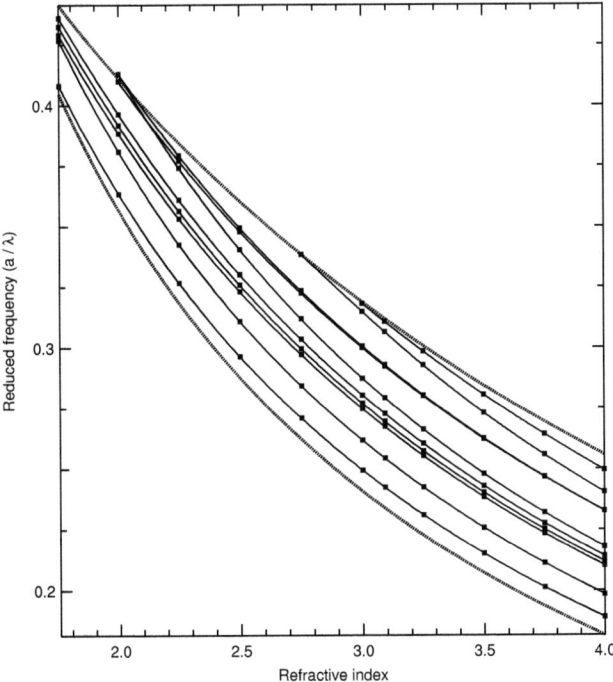

Figure B.5: Reduced energy of the modes of the modified L3 cavity ($d = 0.15 \cdot a$ and $r = \frac{2}{3}r_0$) as a function of the refractive index.

Chapter B. Modes of selected photonic crystal cavities

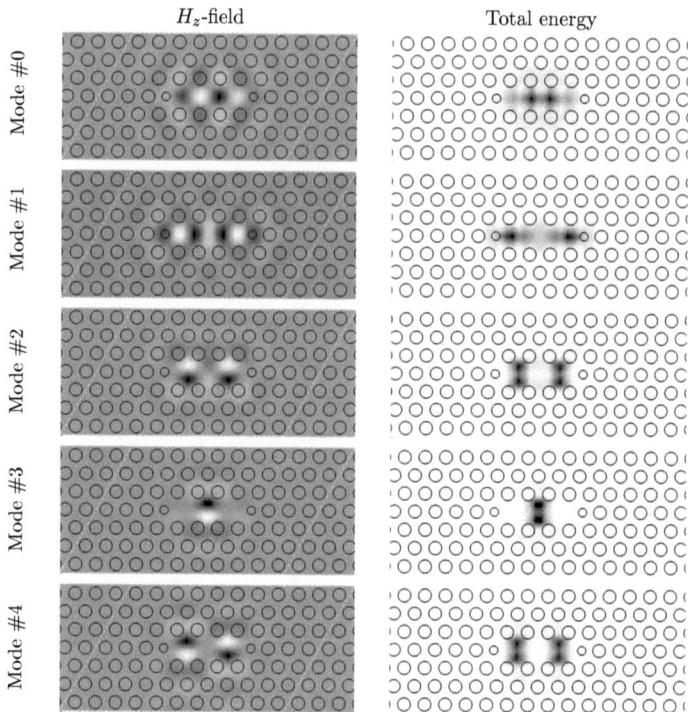

Figure B.6: The modes of the modified L3 cavity for $n_{\text{eff}} = 4.00$.

B.3. Modified L3 cavity

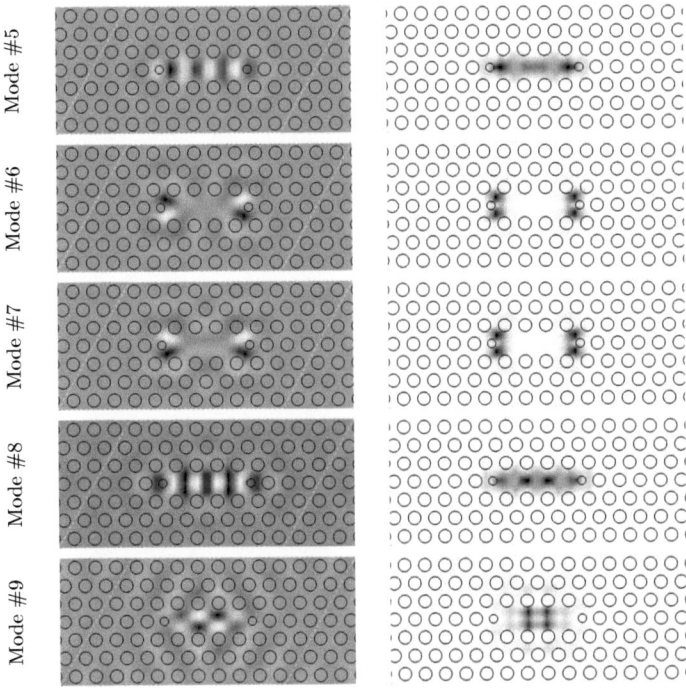

Figure B.7: The modes of the modified L3 cavity for $n_{\text{eff}} = 4.00$. (continued)

Bibliography

[Akahane 03] Y. Akahane, T. Asano, B.-S. Song & S. Noda. *High-Q photonic nanocavity in a two-dimensional photonic crystal.* Nature, vol. 425, page 944, 2003. 30, 31, 47, 53, 76, 104, 110, 111

[Akahane 05] Y. Akahane, T. Asano, B.-S. Song & S. Noda. *Fine-tuned high-Q photonic-crystal nanocavity.* Optics Express, vol. 13, no. 4, page 1202, 2005. 53

[Alloing 05] B. Alloing, C. Zinoni, V. Zwiller, L. H. Li, C. Monat, M. Gobet, G. Buchs, A. Fiore, E. Pelucchi & E. Kapon. *Growth and characterization of single quantum dots emitting at 1300 nm.* Applied Physics Letters, vol. 86, page 101908, 2005. 17, 18, 109

[Alloing 07] B. Alloing, C. Zinoni, L. H. Li, A. Fiore & G. Patriarche. *Structural and optical properties of low-density and In-rich InAs/GaAs quantum dots.* Journal of Applied Physics, vol. 101, no. 2, page 024918, 2007. 17

[Andreani 99] L. C. Andreani, G. Panzarini & J.-M. Gérard. *Strong-coupling regime for quantum boxes in pillar microcavities: Theory.* Physical Review B, vol. 60, no. 19, page 13279, 11 1999. 20

[Anscombe 09] N. Anscombe. *Vienna encrypts communication network.* Optics and laser europe, vol. 168, page 26, 2009. 14

[Badolato 05] A. Badolato, K. Hennessy, M. Atatüre, J. Dreiser, E. Hu, P. M. Petroff & A. Imamo glu. *Deterministic Coupling of Single Quantum Dots to Single Nanocavity Modes.* Science, vol. 308, no. 5725, page 1158, 2005. 21, 48, 111

[Balet 07] L. Balet, M. Francardi, A. Gerardino, N. Chauvin, B. Alloing, C. Zinoni, C. Monat, L. H. Li, N. Le Thomas, R. Houdré & A. Fiore. *Enhanced spontaneous emission rate from single InAs quantum dots in a photonic crystal nanocavity at telecom wavelengths.* Applied Physics Letters, vol. 91, no. 12, page 123115, 2007. 109

[Bayer 01]   M. Bayer, T. L. Reinecke, F. Weidner, A. Larionov, A. McDonald & A. Forchel. *Inhibition and Enhancement of the Spontaneous Emission of Quantum Dots in Structured Microresonators*. Physical Review Letters, vol. 86, no. 14, page 3168, 2001. 20

[Bayer 02a]   M. Bayer & A. Forchel. *Temperature dependence of the exciton homogeneous linewidth in $In_{0.60}Ga_{0.40}As/GaAs$ self-assembled quantum dots*. Physical Review B, vol. 65, no. 4, page 041308, 2002. 17

[Bayer 02b]   M. Bayer, G. Ortner, O. Stern, A. Kuther, A. A. Gorbunov, A. Forchel, P. Hawrylak, S. Fafard, K. Hinzer, T. L. Reinecke, S. N. Walck, J. P. Reithmaier, F. Klopf & F. Schafer. *Fine structure of neutral and charged excitons in self-assembled In(Ga)As/(Al) GaAs quantum dots*. Physical Review B, vol. 65, no. 19, page 195315, 2002. 111

[Benisty 98]   H. Benisty, J.-M. Gérard, R. Houdré, J. Rarity & C. Weisbuch, editeurs. *Confined photon systems - fundamentals and applications*. Springer, 1998. 20

[Bennett 84]   C. H. Bennett & G. Brassard. *Quantum cryptography: Public Key Distribution and Coin Tossing*. Proc. IEEE International Conference on Computers, Systems and Signal Processing (Bangalore, India), 1984. 13

[Beveratos 02]   A. Beveratos, R. Brouri, T. Gacoin, A. Villing, J.-P. Poizat & P. Grangier. *Single Photon Quantum Cryptography*. Physical Review Letters, vol. 89, no. 18, page 187901, 10 2002. 14

[Braive 08]   R. Braive. *Contrôle de l'émission spontanée dans les cavités à cristal photonique*. PhD thesis, Université de Paris 7, 2008. 84

[Burnham 70]   D. C. Burnham & D. L. Weinberg. *Observation of Simultaneity in Parametric Production of Optical Photon Pairs*. Physical Review Letters, vol. 25, no. 2, page 84, 1970. 14

[Chang 06]   W.-H. Chang, W.-Y. Chen, H.-S. Chang, T.-P. Hsieh, J.-I. Chyi & T.-M. Hsu. *Efficient Single-Photon Sources Based on Low-Density Quantum Dots in Photonic-Crystal Nanocavities*. Physical Review Letters, vol. 96, no. 11, page 117401, 2006. 20, 113

[Childress 06] L. Childress, J. M. Taylor, A. S. Sørensen & M. D. Lukin. *Fault-Tolerant Quantum Communication Based on Solid-State Photon Emitters*. Physical Review Letters, vol. 96, no. 7, page 070504, 2006. 14

[Clauser 74] J. F. Clauser. *Experimental distinction between the quantum and classical field-theoretic predictions for the photoelectric effect*. Physical Review D, vol. 9, no. 4, page 853, 1974. 15

[Combrié 08] S. Combrié, A. De Rossi, Q. V. Tran & H. Benisty. *GaAs photonic crystal cavity with ultrahigh Q: microwatt nonlinearity at 1.55 µm*. Optics Letters, vol. 33, no. 16, page 1908, 2008. 40, 48

[Dou 08] X. M. Dou, X. Y. Chang, B. Q. Sun, Y. H. Xiong, Z. C. Niu, S. S. Huang, H. Q. Ni, Y. Du & J. B. Xia. *Single-photon-emitting diode at liquid nitrogen temperature*. Applied Physics Letters, vol. 93, no. 10, page 101107, 2008. 16

[Einstein 05] A. Einstein. *Über einen die Erzeugung und Verwandlung des Lichtes betreffenden heuristischen Gesichtspunkt*. Annalen der Physik, vol. 322, no. 6, page 132, 1905. 15

[Ekert 96] A. Ekert & R. Jozsa. *Quantum computation and Shor's factoring algorithm*. Reviews of Modern Physics, vol. 68, no. 3, page 733, Jul 1996. 13

[El-Kallassi 08] P. El-Kallassi, S. Balog, R. Houdré, L. Balet, L. H. Li, M. Francardi, A. Gerardino, A. Fiore, R. Ferrini & L. Zuppiroli. *Local infiltration of planar photonic crystals with UV-curable polymers*. Journal of the Optical Society of America B, vol. 25, no. 10, page 1562, 2008. 64

[Englund 05] D. Englund, D. Fattal, E. Waks, G. Solomon, B. Zhang, T. Nakaoka, Y. Arakawa, Y. Yamamoto & J. Vučković. *Controlling the Spontaneous Emission Rate of Single Quantum Dots in a Two-Dimensional Photonic Crystal*. Physical Review Letters, vol. 95, no. 1, page 013904, 2005. 20

[Englund 07] D. Englund, A. Faraon, I. Fushman, N. Stoltz, P. Petroff & J. Vučković. *Controlling cavity reflectivity with a single quantum dot*. Nature, vol. 450, no. 7171, page 857, 12 2007. 20, 49

[Fattal 04a] D. Fattal, E. Diamanti, K. Inoue & Y. Yamamoto. *Quantum Teleportation with a Quantum Dot Single Photon Source*. Physical Review Letters, vol. 92, no. 3, page 037904, 2004. 13

[Fattal 04b] D. Fattal, K. Inoue, J. Vučković, C. Santori, G. S. Solomon & Y. Yamamoto. *Entanglement Formation and Violation of Bell's Inequality with a Semiconductor Single Photon Source*. Physical Review Letters, vol. 92, no. 3, page 037903, 2004. 13

[Ferrini 02] R. Ferrini, D. Leuenberger, M. Mulot, Q. Min, R. Moosburger, M. Kamp, A. Forchel, S. Anand & R. Houdré. *Optical study of two-dimensional InP-based photonic crystals by internal light source technique*. IEEE Journal of Quantum Electronics, vol. 38, no. 7, page 786, 2002. 53

[Francardi 08] M. Francardi, L. Balet, A. Gerardino, N. Chauvin, D. Bitauld, L. H. Li, B. Alloing & A. Fiore. *Enhanced spontaneous emission in a photonic-crystal light-emitting diode*. Applied Physics Letters, vol. 93, no. 14, page 143102, 2008. 115

[Fry 00] P. W. Fry, I. E. Itskevich, D. J. Mowbray, M. S. Skolnick, J. J. Finley, J. A. Barker, E. P. O'Reilly, L. R. Wilson, I. A. Larkin, P. A. Maksym, M. Hopkinson, M. Al-Khafaji, J. P. R. David, A. G. Cullis, G. Hill & J. C. Clark. *Inverted Electron-Hole Alignment in InAs-GaAs Self-Assembled Quantum Dots*. Physical Review Letters, vol. 84, no. 4, 2000. 96

[Fushman 07] I. Fushman, E. Waks, D. Englund, N. Stoltz, P. Petroff & J. Vučković. *Ultrafast nonlinear optical tuning of photonic crystal cavities*. Applied Physics Letters, vol. 90, no. 9, page 091118, 2007. 49

[Galli 09] M. Galli, S. L. Portalupi, M. Belotti, L. C. Andreani, L. O'Faolain & T. F. Krauss. *Light scattering and Fano resonances in high-Q photonic crystal nanocavities*. Applied Physics Letters, vol. 94, no. 7, page 071101, 2009. 47

[Gallo 08] P. Gallo, M. Felici, B. Dwir, K. A. Atlasov, K. F. Karlsson, A. Rudra, A. Mohan, G. Biasiol, L. Sorba & E. Kapon. *Integration of site-controlled pyramidal quantum dots and photonic crystal membrane cavities*. Applied Physics Letters, vol. 92, no. 26, page 263101, 2008. 48

[Garrigues 02] M. Garrigues, J. L. Leclercq & P. Viktorovitch. *III-V Semiconductor based MOEMS devices for optical telecommunications*. Microelectronic Engineering, vol. 61-62, page 933, 2002. 92

[Gayral 01] B. Gayral, J. M. Gérard, B. Sermage, A. Lemaitre & C. Dupuis. *Time-resolved probing of the Purcell effect for InAs quantum boxes in GaAs microdisks*. Applied Physics Letters, vol. 78, no. 19, page 2828, 2001. 20, 33, 113

[Gayral 08] B. Gayral & J. M. Gérard. *Photoluminescence experiment on quantum dots embedded in a large Purcell-factor microcavity*. Physical Review B, vol. 78, no. 23, page 235306, 2008. 114

[Gérard 98] J. M. Gérard, B. Sermage, B. Gayral, B. Legrand, E. Costard & V. Thierry-Mieg. *Enhanced Spontaneous Emission by Quantum Boxes in a Monolithic Optical Microcavity*. Physical Review Letters, vol. 81, no. 5, page 1110, 1998. 20, 33

[Gérard 99] J.-M. Gérard & B. Gayral. *Strong Purcell effect for InAs quantum boxes in three-dimensional solid-state microcavities*. Journal of Lightwave Technology, vol. 17, no. 11, page 2089, 1999. 20

[Gérard 01] J. M. Gérard & B. Gayral. *InAs quantum dots: artificial atoms for solid-state cavity-quantum electrodynamics*. Physica E: Low-dimensional Systems and Nanostructures, vol. 9, no. 1, page 131, 2001. 20

[Gevaux 06] D. G. Gevaux, A. J. Bennett, R. M. Stevenson, A. J. Shields, P. Atkinson, J. Griffiths, D. Anderson, G. A. C. Jones & D. A. Ritchie. *Enhancement and suppression of spontaneous emission by temperature tuning InAs quantum dots to photonic crystal cavities*. Applied Physics Letters, vol. 88, no. 13, page 131101, 2006. 20, 113

[Gisin 02] N. Gisin, G. Ribordy, W. Tittel & H. Zbinden. *Quantum cryptography*. Reviews of Modern Physics, vol. 74, no. 1, pages 145–195, Mar 2002. 14

[Gol'tsman 01] G. N. Gol'tsman, O. Okunev, G. Chulkova, A. Lipatov, A. Semenov, K. Smirnov, B. Voronov, A. Dzardanov, C. Williams & R. Sobolewski. *Pi-

cosecond superconducting single-photon optical detector. Applied Physics Letters, vol. 79, no. 6, page 705, 2001. 44

[Graham 99] L. A. Graham, D. L. Huffaker & D. G. Deppe. *Spontaneous lifetime control in a native-oxide-apertured microcavity*. Applied Physics Letters, vol. 74, no. 17, page 2408, 1999. 20, 33

[Happ 02] T. D. Happ, I. I. Tartakovskii, V. D. Kulakovskii, J.-P. Reithmaier, M. Kamp & A. Forchel. *Enhanced light emission of $In_x Ga_{1-x} As$ quantum dots in a two-dimensional photonic-crystal defect microcavity*. Physical Review B, vol. 66, no. 4, page 041303, 7 2002. 33

[Hennessy 05] K. Hennessy, A. Badolato, A. Tamboli, P. M. Petroff, E. Hu, M. Atatüre, J. Dreiser & A. Imamoğlu. *Tuning photonic crystal nanocavity modes by wet chemical digital etching*. Applied Physics Letters, vol. 87, no. 2, page 021108, 2005. 48

[Hennessy 06] K. Hennessy, C. Hogerle, E. Hu, A. Badolato & A. Imamoğlu. *Tuning photonic nanocavities by atomic force microscope nano-oxidation*. Applied Physics Letters, vol. 89, page 041118, 2006. 48

[Hennessy 07] K. Hennessy, A. Badolato, M. Winger, D. Gerace, M. Atatüre, S. Gulde, S. Fält, E. L. Hu & A. Imamoğlu. *Quantum nature of a strongly coupled single quantum dot–cavity system*. Nature, vol. 445, page 896, 2007. 20, 24, 48, 110

[Herrmann 06] R. Herrmann, T. Sünner, T. Hein, A. Löffler, M. Kamp & A. Forchel. *Ultrahigh-quality photonic crystal cavity in GaAs*. Optics Letters, vol. 31, no. 9, pages 1229–1231, 2006. 48

[Hillmer 89] H. Hillmer, A. Forchel, S. Hansmann, M. Morohashi, E. Lopez, H. P. Meier & K. Ploog. *Optical investigations on the mobility of two-dimensional excitons in $GaAs/Ga_{1-x}Al_x As$ quantum wells*. Physical Review B, vol. 39, no. 15, page 10901, 1989. 120

[Hofbauer 07] F. Hofbauer, S. Grimminger, J. Angele, G. Bohm, R. Meyer, M. C. Amann & J. J. Finley. *Electrically probing photonic bandgap phenomena in contacted*

	defect nanocavities. Applied Physics Letters, vol. 91, no. 20, page 201111, 2007. 116
[Högele 08]	A. Högele, C. Galland, M. Winger & A. Imamoğlu. *Photon Antibunching in the Photoluminescence Spectra of a Single Carbon Nanotube.* Physical Review Letters, vol. 100, no. 21, page 217401, 2008. 16
[Hong 86]	C. K. Hong & L. Mandel. *Experimental realization of a localized one-photon state.* Physical Review Letters, vol. 56, no. 1, page 58, 1986. 14
[Hopman 06]	W. C. L. Hopman, K. O. van der Werf, A. J. F. Hollink, W. Bogaerts, V. Subramaniam & R. M. de Ridder. *Nano-mechanical tuning and imaging of a photonic crystal micro-cavity resonance.* Optics Express, vol. 14, no. 19, page 8745, 2006. 49, 75
[Hwang 03]	W.-Y. Hwang. *Quantum Key Distribution with High Loss: Toward Global Secure Communication.* Physical Review Letters, vol. 91, no. 5, page 057901, 2003. 14
[Imamoğlu 99]	A. Imamoğlu & Y. Yamamoto. Mesoscopic quantum optics. Wiley-Intersicence, 1999. 23
[Intonti 06]	F. Intonti, S. Vignolini, V. Turck, M. Colocci, P. Bettotti, L. Pavesi, S. L. Schweizer, R. Wehrspohn & D. Wiersma. *Rewritable photonic circuits.* Applied Physics Letters, vol. 89, page 211117, 2006. 48, 64, 67
[Intonti 08a]	F. Intonti, S. Vignolini, F. Riboli, A. Vinattieri, D. S. Wiersma, M. Colocci, L. Balet, C. Monat, C. Zinoni, L. H. Li, R. Houdré, M. Francardi, A. Gerardino, A. Fiore & M. Gurioli. *Spectral tuning and near-field imaging of photonic crystal microcavities.* Physical Review B, vol. 78, no. 4, page 041401, 2008. 68, 71, 73, 75
[Intonti 08b]	F. Intonti, S. Vignolini, F. Riboli, A. Vinattieri, D. S. Wiersma, M. Colocci, M. Gurioli, L. Balet, C. Monat, L. H. Li, N. Le Thomas, R. Houdré, A. Fiore, M. Francardi, A. Gerardino, F. Römer & B. Witzigmann. *Near-field mapping of quantum dot emission from single-photonic crystal cavity modes.* Physica E: Low-dimensional Systems and Nanostructures, vol. 40, no. 6, page 1965, 2008. 71, 73

[Joannopoulos 95] J. D. Joannopoulos, R. D. Meade & J. N. Winn. Photonic crystals: Molding the flow of light. Princeton University Press, Princeton, NJ, 1995. 25, 125

[Kamada 08] H. Kamada & T. Kutsuwa. *Broadening of single quantum dot exciton luminescence spectra due to interaction with randomly fluctuating environmental charges*. Physical Review B, vol. 78, page 155324, 2008. 18, 110

[Khitrova 06] G. Khitrova, H. M. Gibbs, M. Kira, S. W. Koch & A. Scherer. *Vacuum Rabi splitting in semiconductors*. Nature Physics, vol. 2, no. 2, page 81, 02 2006. 20

[Kim 06] S.-H. Kim, S.-K. Kim & Y.-H. Lee. *Vertical beaming of wavelength-scale photonic crystal resonators*. Physical Review B, vol. 73, no. 23, page 235117, 2006. 101

[Kimble 77] H. J. Kimble, M. Dagenais & L. Mandel. *Photon Antibunching in Resonance Fluorescence*. Physical Review Letters, vol. 39, no. 11, page 691, 1977. 15

[Kiraz 01] A. Kiraz, P. Michler, C. Becher, B. Gayral, A. Imamoğlu, Lidong Zhang, E. Hu, W. V. Schoenfeld & P. M. Petroff. *Cavity-quantum electrodynamics using a single InAs quantum dot in a microdisk structure*. Applied Physics Letters, vol. 78, no. 25, page 3932, 2001. 20, 33

[Kittel 04] C. Kittel. Introduction to solid state physics. Wiley, New York, 2004. 27

[Klimov 04] V. I. Klimov. Semiconductor and metal nanocrystals. Marcel Dekker, Inc, 2004. 15

[Knill 01] E. Knill, R. Laflamme & G. J. Milburn. *A scheme for efficient quantum computation with linear optics*. Nature, vol. 409, no. 6816, page 46, 2001. 13

[Koenderink 05] A. F. Koenderink, M. Kafesaki, B. C. Buchler & V. Sandoghdar. *Controlling the Resonance of a Photonic Crystal Microcavity by a Near-Field Probe*. Physical Review Letters, vol. 95, no. 15, page 153904, 2005. 70

[Korneev 07] A. Korneev, Y. Vachtomin, O. Minaeva, A. Divochiy, K. Smirnov, O. Okunev, G. Gol'tsman, C. Zinoni, N. Chauvin, L. Balet, F. Marsili, D. Bitauld, B. Alloing, Lianhe Li, A. Fiore, L. Lunghi, A. Gerardino, M. Halder, C. Jorel &

BIBLIOGRAPHY

H. Zbinden. *Single-Photon Detection System for Quantum Optics Applications*. Selected Topics in Quantum Electronics, IEEE Journal of, vol. 13, no. 4, pages 944–951, 2007. 44

[Kress 05]   A. Kress, F. Hofbauer, N. Reinelt, M. Kaniber, H. J. Krenner, R. Meyer, G. Bohm & J. J. Finley. *Manipulation of the spontaneous emission dynamics of quantum dots in two-dimensional photonic crystals*. Physical Review B, vol. 71, no. 24, page 241304, 2005. 20

[Kuramochi 06]   E. Kuramochi, M. Notomi, S. Mitsugi, A. Shinya, T. Tanabe & T. Watanabe. *Ultrahigh-Q photonic crystal nanocavities realized by the local width modulation of a line defect*. Applied Physics Letters, vol. 88, no. 4, page 041112, 2006. 47, 55

[Lalouat 07]   L. Lalouat, B. Cluzel, P. Velha, E. Picard, D. Peyrade, J. P. Hugonin, P. Lalanne, E. Hadji & F. de Fornel. *Near-field interactions between a subwavelength tip and a small-volume photonic-crystal nanocavity*. Physical Review B, vol. 76, page 041102, 2007. 75

[Leonard 00]   S. W. Leonard, J. P. Mondia, H. M. van Driel, O. Toader, S. John, K. Busch, A. Birner, U. Gösele & V. Lehmann. *Tunable two-dimensional photonic crystals using liquid crystal infiltration*. Physical Review B, vol. 61, page R2389, 2000. 49

[Loudon 00]   R. Loudon. *The quantum theory of light*. Oxford Universtiy Press, third edition, 2000. 18, 22

[Lounis 00a]   B. Lounis, H. A. Bechtel, D. Gerion, P. Alivisatos & W. E. Moerner. *Photon antibunching in single CdSe/ZnS quantum dot fluorescence*. Chemical Physics Letters, vol. 329, no. 5, page 399, 2000. 15

[Lounis 00b]   B. Lounis & W. E. Moerner. *Single photons on demand from a single molecule at room temperature*. Nature, vol. 407, no. 6803, page 491, 2000. 15

[Lounis 05]   B. Lounis & M. Orrit. *Single-photon sources*. Reports on Progress in Physics, vol. 68, no. 5, page 1129, 2005. 12

[Märki 06]   I. Märki, M. Salt & H. P. Herzig. *Tuning the resonance of a photonic crystal microcavity with an AFM probe*. Optics Express, vol. 14, page 2969, 2006. 49, 75

[Marple 64]   D. T. F. Marple. *Refractive Index of GaAs*. Journal of Applied Physics, vol. 35, no. 4, page 1241, 1964. 78

[Michler 00a]   P. Michler, A. Imamoğlu, M. D. Mason, P. J. Carson, G. F. Strouse & S. K. Buratto. *Quantum correlation among photons from a single quantum dot at room temperature*. Nature, vol. 406, no. 6799, page 968, 2000. 15

[Michler 00b]   P. Michler, A. Kiraz, C. Becher, W. V. Schoenfeld, P. M. Petroff, Lidong Zhang, E. Hu & A. Imamoğlu. *A Quantum Dot Single-Photon Turnstile Device*. Science, vol. 290, no. 5500, page 2282, 2000. 16

[Mingaleev 04]   S. F. Mingaleev, M. Schillinger, D. Hermann & K. Busch. *Tunable photonic crystal circuits: concepts and designs based on single-pore infiltration*. Optics Letters, vol. 29, no. 24, page 2858, 2004. 67

[Mirin 04]   R. P. Mirin. *Photon antibunching at high temperature from a single In-GaAs/GaAs quantum dot*. Applied Physics Letters, vol. 84, no. 8, page 1260, 2004. 16

[Moreau 01]   E. Moreau, I. Robert, J. M. Gérard, I. Abram, L. Manin & V. Thierry-Mieg. *Single-mode solid-state single photon source based on isolated quantum dots in pillar microcavities*. Applied Physics Letters, vol. 79, no. 18, page 2865, 2001. 33

[Mosor 05]   S. Mosor, J. Hendrickson, B. C. Richards, J. Sweet, G. Khitrova, H. M. Gibbs, T. Yoshie, A. Scherer, O. B. Shchekin & D. G. Deppe. *Scanning a photonic crystal slab nanocavity by condensation of xenon*. Applied Physics Letters, vol. 87, no. 14, page 141105, 2005. 49, 59, 61

[Mujumdar 07]   S. Mujumdar, A. F. Koenderink, T. Sünner, B. C. Buchler, M. Kamp, A. Forchel & V. Sandoghdar. *Near-field imaging and frequency tuning of a high-Q photonic crystal membrane microcavity*. Optics Express, vol. 15, no. 25, page 17214, 2007. 49, 75

[Nielsen 00] M. A. Nielsen & I. L. Chuang. *Quantum computation and quantum information*. Cambridge University Press, 2000. 12

[Noda 07] S. Noda, M. Fujita & T. Asano. *Spontaneous-emission control by photonic crystals and nanocavities*. Nature Photonics, vol. 1, no. 8, page 449, 08 2007. 47

[Nomura 06] M. Nomura, S. Iwamoto, M. Nishioka, S. Ishida & Y. Arakawa. *Highly efficient optical pumping of photonic crystal nanocavity lasers using cavity resonant excitation*. Applied Physics Letters, vol. 89, no. 16, page 161111, 2006. 94

[Notomi 06] M. Notomi, H. Taniyama, S. Mitsugi & E. Kuramochi. *Optomechanical Wavelength and Energy Conversion in High-Q Double-Layer Cavities of Photonic Crystal Slabs*. Physical Review Letters, vol. 97, no. 2, page 023903, 2006. 75, 76

[Notomi 07] M. Notomi, T. Tanabe, A. Shinya, E. Kuramochi, H. Taniyama, S. Mitsugi & M. Morita. *Nonlinear and adiabatic control of high-Qphotonic crystal nanocavities*. Optics Express, vol. 15, no. 26, page 17458, 2007. 47

[O'Brien 03] J. L. O'Brien, G. J. Pryde, A. G. White, T. C. Ralph & D. Branning. *Demonstration of an all-optical quantum controlled-NOT gate*. Nature, vol. 426, no. 6964, page 264, 2003. 13

[Park 04] H.-G. Park, S.-H. Kim, S.-H. Kwon, Y.-G. Ju, J.-K. Yang, J.-H. Baek, S.-B. Kim & Y.-H. Lee. *Electrically Driven Single-Cell Photonic Crystal Laser*. Science, vol. 305, no. 5689, page 1444, 2004. 116

[Pelton 02] M. Pelton, C. Santori, J. Vučković, B. Zhang, G. S. Solomon, J. Plant & Y. Yamamoto. *Efficient Source of Single Photons: A Single Quantum Dot in a Micropost Microcavity*. Physical Review Letters, vol. 89, no. 23, page 233602, Nov 2002. 33

[Peter 05] E. Peter, P. Senellart, D. Martrou, A. Lemaitre, J. Hours, J. M. Gérard & J. Bloch. *Exciton-Photon Strong-Coupling Regime for a Single Quantum Dot Embedded in a Microcavity*. Physical Review Letters, vol. 95, no. 6, page 067401, 2005. 20

BIBLIOGRAPHY

[Pimpinelli 98] A. Pimpinelli. *Physics of crystal growth.* Cambridge University Press, 1998. 16

[Polzik 92] E. S. Polzik, J. Carri & H. J. Kimble. *Spectroscopy with squeezed light.* Physical Review Letters, vol. 68, no. 20, page 3020, 1992. 12

[Purcell 46] E. M. Purcell. *Spontaneous Emission Probabilities at Radio Frequencies.* Physical Review, vol. 69, page 681, 1946. 20

[Quantis 04] Quantis. *Random Numbers Generation using Quantum Physics.* white paper, id Quantique SA, 2004. 12

[Rarity 94] J. G. Rarity, P. C. M. Owens & P. R. Tapster. *Quantum Random-number Generation and Key Sharing.* Journal of Modern Optics, vol. 41, no. 12, page 2435, 1994. 12

[Reese 01] C. Reese, C. Becher, A. Imamoğlu, E. Hu, B. D. Gerardot & P. M. Petroff. *Photonic crystal microcavities with self-assembled InAs quantum dots as active emitters.* Applied Physics Letters, vol. 78, no. 16, page 2279, 2001. 33

[Reithmaier 04] J. P. Reithmaier, G. Sek, A. Loffler, C. Hofmann, S. Kuhn, S. Reitzenstein, L. V. Keldysh, V. D. Kulakovskii, T. L. Reinecke & A. Forchel. *Strong coupling in a single quantum dot-semiconductor microcavity system.* Nature, vol. 432, no. 7014, page 197, 11 2004. 20, 33

[Rigneault 01] H. Rigneault, J. Broudic, B. Gayral & J.-M. Gérard. *Far-field radiation from quantum boxes located in pillar microcavities.* Optics Letters, vol. 26, no. 20, page 15957, 2001. 97

[Römer 07a] F. Römer & B. Witzigmann. *Investigation of the optical farfield of photonic crystal microcavities.* Proceedings of SPIE, vol. 6480, page 64801B, 2007. 98

[Römer 07b] F. Römer, B. Witzigmann, O. Chinellato & P. Arbenz. *Investigation of the Purcell effect in photonic crystal cavities with a 3D Finite Element Maxwell Solver.* Optical and Quantum Electronics, vol. 39, no. 4, page 341, 03 2007. 27, 30

BIBLIOGRAPHY

[Rosa 05] L. Rosa, S. Selleri & F. Poli. *Design of photonic-crystal and wire waveguide interface*. Journal of Lightwave Technology, vol. 23, no. 9, page 2740, 9 2005. 105

[Rosencher 02] Emmanuel Rosencher & Borge Vinter. Optoelectronics. Cambridge University Press, 2002. 78, 94

[Ryu 03] H. Y. Ryu & M. Notomi. *Enhancement of spontaneous emission from the resonant modes of a photonic crystal slab single-defect cavity*. Optics Letters, vol. 28, no. 23, page 2390, 2003. 24

[Saleh 91] B. E. A. Saleh & M. C. Teich. Fundamenteals of photonics. Wiley, 1 edition, 1991. 78

[Santori 01] C. Santori, M. Pelton, G. Solomon, Y. Dale & Y. Yamamoto. *Triggered Single Photons from a Quantum Dot*. Physical Review Letters, vol. 86, no. 8, page 1502, 2001. 16

[Shields 07] A. J. Shields. *Semiconductor quantum light sources*. Nature Photonics, vol. 1, no. 4, page 215, 04 2007. 12

[Shimizu 01] K. T. Shimizu, R. G. Neuhauser, C. A. Leatherdale, S. A. Empedocles, W. K. Woo & M. G. Bawendi. *Blinking statistics in single semiconductor nanocrystal quantum dots*. Physical Review B, vol. 63, no. 20, page 205316, 2001. 15

[Solomon 01] G. S. Solomon, M. Pelton & Y. Yamamoto. *Single-mode Spontaneous Emission from a Single Quantum Dot in a Three-Dimensional Microcavity*. Physical Review Letters, vol. 86, no. 17, page 3903, 2001. 20, 33

[Song 05] B.-S. Song, S. Noda, T. Asano & Y. Akahane. *Ultra-high-Q photonic double-heterostructure nanocavity*. Nature Materials, vol. 4, no. 3, page 207, 2005. 47

[Song 07] B.-S. Song, T. Asano & S. Noda. *Heterostructures in two-dimensional photonic-crystal slabs and their application to nanocavities*. Journal of Physics D, vol. 40, no. 9, pages 2629–2634, 2007. 47

[Srinivasan 02] K. Srinivasan & O. Painter. *Momentum space design of high-Q photonic crystal optical cavities.* Optics Express, vol. 10, page 670, 2002. 29, 76

[Srinivasan 07] K. Srinivasan & O. Painter. *Linear and nonlinear optical spectroscopy of a strongly coupled microdisk-quantum dot system.* Nature, vol. 450, no. 7171, page 862, 12 2007. 20

[Stranski 39] I. N. Stranski & L. von Krastanow. *Abhandlungen der Mathematisch-Naturwissenschaftlichen Klasse.* Akademie der Wissenschaften und der Literatur in Mainz, vol. 146, page 797, 1939. 16

[Tanabe 07] T. Tanabe, A. Shinya, E. Kuramochi, S. Kondo, H. Taniyama & M. Notomi. *Single point defect photonic crystal nanocavity with ultrahigh quality factor achieved by using hexapole mode.* Applied Physics Letters, vol. 91, no. 2, page 021110, 2007. 47

[Tanaka 08] Y. Tanaka, T. Asano & S. Noda. *Design of Photonic Crystal Nanocavity With Q-Factor of $\sim 10^9$.* Journal of Lightwave Technology, vol. 26, no. 11, page 1532, 2008. 47

[Thompson 01] R. M. Thompson, R. M. Stevenson, A. J. Shields, I. Farrer, C. J. Lobo, D. A. Ritchie, M. L. Leadbeater & M. Pepper. *Single-photon emission from exciton complexes in individual quantum dots.* Physical Review B, vol. 64, no. 20, page 201302, 2001. 16

[Topolancik 07] J. Topolancik, F. Vollmer & B. Ilic. *Random high-Q cavities in disordered photonic crystal waveguides.* Applied Physics Letters, vol. 91, no. 20, page 201102, 2007. 47

[Vaccaro 01] P. O. Vaccaro, K. Kubota & T. Aida. *Strain-driven self-positioning of micromachined structures.* Applied Physics Letters, vol. 78, no. 19, page 2852, 2001. 93

[Vahala 03] K. J. Vahala. *Optical microcavities.* Nature, vol. 424, no. 6950, page 839, 2003. 20

[Vignolini 08] S. Vignolini, F. Intonti, L. Balet, M. Zani, F. Riboli, A. Vinattieri, D. S. Wiersma, M. Colocci, L. H. Li, M. Francardi, A. Gerardino, A. Fiore &

M. Gurioli. *Nonlinear optical tuning of photonic crystal microcavities by near-field probe.* Applied Physics Letters, vol. 93, page 023124, 2008. 72, 74

[Vignolini 09] S. Vignolini, F. Intonti, M. Zani, F. Riboli, D. S. Wiersma, L. H. Li, L. Balet, M. Francardi, A. Gerardino, A. Fiore & M. Gurioli. *Near-field imaging of coupled photonic-crystal microcavities.* Applied Physics Letters, vol. 94, no. 15, page 151103, 2009. 53, 75

[Vučković 03] J. Vučković, D. Fattal, C. Santori, G. S. Solomon & Y. Yamamoto. *Enhanced single-photon emission from a quantum dot in a micropost microcavity.* Applied Physics Letters, vol. 82, no. 21, page 3596, 2003. 33

[Waks 02] E. Waks, K. Inoue, C. Santori, D. Fattal, J. Vučković, G. S. Solomon & Y. Yamamoto. *Secure communication: Quantum cryptography with a photon turnstile.* Nature, vol. 420, no. 6917, page 762, 12 2002. 14

[Winger 08] M. Winger, A. Badolato, K. J. Hennessy, E. L. Hu & A. Imamoğlu. *Quantum Dot Spectroscopy Using Cavity Quantum Electrodynamics.* Physical Review Letters, vol. 101, no. 22, page 226808, 2008. 20

[Wootters 82] W. K. Wootters & W. H. Zurek. *A single quantum cannot be cloned.* Nature, vol. 299, no. 5886, page 802, 10 1982. 13

[Xiao 87] M. Xiao, L.-A. Wu & H. J. Kimble. *Precision measurement beyond the shot-noise limit.* Physical Review Letters, vol. 59, no. 3, page 278, 1987. 12

[Yoshie 01] T. Yoshie, A. Scherer, H. Chen, D. Huffaker & D. Deppe. *Optical characterization of two-dimensional photonic crystal cavities with indium arsenide quantum dot emitters.* Applied Physics Letters, vol. 79, no. 1, page 114, 2001. 33

[Yoshie 04] T. Yoshie, A. Scherer, J. Hendrickson, G. Khitrova, H. M. Gibbs, G. Rupper, C. Ell, O. B. Shchekin & D. G. Deppe. *Vacuum Rabi splitting with a single quantum dot in a photonic crystal nanocavity.* Nature, vol. 432, no. 7014, pages 200–203, 2004. 20, 33, 49

[Yuan 02] Z. Yuan, B. E. Kardynal, R. M. Stevenson, A. J. Shields, C. J. Lobo, K. Cooper, N. S. Beattie, D. A. Ritchie & M. Pepper. *Electrically Driven Single-Photon Source.* Science, vol. 295, no. 5552, page 102, 2002. 16

[Zinoni 06] C. Zinoni, B. Alloing, C. Monat, V. Zwiller, L. H. Li, A. Fiore, L. Lunghi, A. Gerardino, H. de Riedmatten, H. Zbinden & N. Gisin. *Time-resolved and antibunching experiments on single quantum dots at 1300 nm.* Applied Physics Letters, vol. 88, no. 13, page 131102, 2006. 18, 113, 121

[Zinoni 07] C. Zinoni, B. Alloing, L. H. Li, F. Marsili, A. Fiore, L. Lunghi, A. Gerardino, Yu. B. Vakhtomin, K. V. Smirnov & G. N. Gol'tsman. *Single-photon experiments at telecommunication wavelengths using nanowire superconducting detectors.* Applied Physics Letters, vol. 91, no. 3, page 031106, 2007. 19

Acknowledgments

During the last four years, I had the opportunity to meet a lot of interesting persons who, directly or not, contributed to the work documented in this volume.

I am especially grateful to my thesis advisor, **Andrea Fiore** for allowing me to join his group. Thank you for always taking the time for discussions, for channeling my enthusiasm and for showing interest in my work.

Nicolas Chauvin and **David Bitauld**, not only for the countless hours spent in a dark laboratory measuring single photons, but also, together with **Francesco Marsili**, for being my fellows in Lausanne and Eindhoven.

Pablo Moreno for welcoming me in his office in Lausanne and for the biking excursions in the late Friday afternoons. **Blandine Alloing** for your wonderful QD samples and for always being cheerful and supportive.

Matthias Skacel and **Saeedeh Jahanmiri Nejad**, good luck with your projects in Eindhoven!

Many thanks to our administrative personal, **Aline Gruaz**, **Pierrette Paulou**, **Claire-Lyse Rouiller**, and **Oana Chiper** in Lausanne. A special "dank je well" to **Margriet van Doorne** for patiently distilling the Dutch way of life into my Swiss habits.

I'm also really grateful to our technical staff for their outstanding work. In particular **Jos van Ruijven**, **Rian Hamhuis**, **Barry Smalbrugge**, **Nicolas Leiser**, **Damien Troillet**, and **Yoan Troillet**.

I also would like to thank **Polina, Marc, Sergei, Ilya,** and **Victor** who encouraged me to start the PhD adventure. Thank you also to the members of the institute, of the Quantum Device group, of PSN, of the International Joyful Choir, of the club montagne, flat-mates, tenors of the Novantica and so much more: **Carl, Christelle, Philipp, Alex, Marco, Lianhe, Annamaria, Marco, Niek, Joris, Murat, Jens, Harm, Paul, Samuel, Salman, Andrei, Nut, Ineke, Andrea, Kirill, Dobri, Pascale, Jean, Samuel, Pierre, Aurélie, Sébastien, Eliane, Marie-Paule, Thierry, Oliver, Evelyn, Swati, Jana, Roland, Verena, Benoît, Etienne, Catherine, Pascal, Jan, Helen, Valentin, Karin, Bernhard, Verena, Peter, Heike, Christian, Annika,**

Torre, Anne, Moritz, Sarah, Filippo, Carine, Myriam, Marc, Claudio, Marco, Sri, Irwan, Greg, Christian, Guy-Claude, Stéphane, Jean-Paul,Pierre, Lucas...

I am also grateful to he members of the jury, **Benoît Deveaud Plédran, Jean-Michel Gérard, Pierre Viktorovitch,** and **Olivier Martin,** who took the time to read and discuss my manuscript and gave me the opportunity to present this work.

Thank you **Evi**, for your daily support even 1000 km away, for your excursions ideas that change my mind and for your positive thinking.

My parents, **Georges-André, Tanja, Lars,** and **Nicolas** for their love, advices and constant support.

Die VDM Verlagsservicegesellschaft sucht für wissenschaftliche Verlage abgeschlossene und herausragende

Dissertationen, Habilitationen, Diplomarbeiten, Master Theses, Magisterarbeiten usw.

für die kostenlose Publikation als Fachbuch.

Sie verfügen über eine Arbeit, die hohen inhaltlichen und formalen Ansprüchen genügt, und haben Interesse an einer honorarvergüteten Publikation?

Dann senden Sie bitte erste Informationen über sich und Ihre Arbeit per Email an *info@vdm-vsg.de*.

Sie erhalten kurzfristig unser Feedback!

VDM Verlagsservicegesellschaft mbH
Dudweiler Landstr. 99
D - 66123 Saarbrücken

Telefon +49 681 3720 174
Fax   +49 681 3720 1749

www.vdm-vsg.de

Die VDM Verlagsservicegesellschaft mbH vertritt

Printed by Books on Demand GmbH, Norderstedt / Germany